The Fundamental Connection between Nature and Nurture

A Review of the Evidence

Edited by
Walter R. Gove
G. Russell Carpenter
Vanderbilt University

LexingtonBooks
D.C. Heath and Company
Lexington, Massachusetts
Toronto

Library of Congress Cataloging in Publication Data

Main entry under title:

The Fundamental connection between nature and nurture.

Includes index.
1. Nature and nurture—Addresses, essays, lectures. 2. Mind and body—Addresses, essays, lectures. 3. Human behavior—Addresses, essays, lectures. I. Gove, Walter R. II. Carpenter, G. Russell.
HM106.F85   302   80-8961
ISBN 0-669-04483-0   AACR2

Copyright © 1982 by D.C. Heath and Company

Published simultaneously in Canada

Printed in the United States of America

International Standard Book Number: 0-669-04483-0

Library of Congress Catalog Card Number: 80-8961

Contents

1 Introduction

Walter R. Gove and
G. Russell Carpenter

The chapters that compose this book represent an effort to force social scientists, particularly sociologists, to question one of their key assumptions about human nature. Our intent is to make it very clear that the belief that the human mind and body are separate entities—a belief that has been part of intellectual thought at least back to Descartes—is no longer a tenable position. This book attempts to demonstrate the necessity of treating human behavior as a unity, a biopsychosocial phenomenon, through a review of empirical evidence demonstrating that such an integrated approach to the study of human behavior is essential. Although many themes are emphasized in the various contributions, the book does not attempt a general theoretical synthesis, but points to the necessity for such a theoretical synthesis. It is our opinion that we do not as yet understand human nature sufficiently to develop a theoretical synthesis that would be of much use. At the same time, we feel we have reached the point where it is clear that such a synthesis is needed; at its most general level, this book attempts to show that the time has come to begin laying the groundwork for such a synthesis.

One issue that needs to be very clear in the reader's mind is that this book does *not* fit into the school of thought that has developed around Edward O. Wilson and that is commonly referred to as *sociobiology.* It is our position that sociobiology, as it has become commonly understood, offers some valid insights, but has had the overall effect of retarding the development of the integration of the biological with social and psychological processes. Sociobiology is sufficiently flawed that most social scientists when confronted with it can readily find sufficient grounds to dismiss the perspective, particularly as it runs directly counter to their philosophical predilections.

One of the premises that underlies the present endeavor is that social science in general, and particularly sociology, has reached a point of fragmentation and stagnation such that the discipline, as a discipline, simply is no longer progressing. It is easy to find sources that support this view, although some of the authorities see this as a temporary phase, while others view it as a sign the discipline is becoming moribund.

Blalock (1979) states in his presidential address to the American Sociological Association: "One particularly disappointing feature of our disci-

1

pline is that we have not had the productive interplay between theory and research called for so eloquently by Merton several decades ago'' (p. 881). Blalock sees sociology as bogged down by an inability "to grapple with a number of extremely complex problems" that are both theoretical and methodological, and he calls for a concerted collaborative effort by sociologists to solve these problems. He says, "we can ill afford to go off in our own direction continuing to proliferate fields of specializing, changing our vocabulary whenever we see fit, or merely hoping that somehow or other the products of our miscellaneous studies will add up" (p. 893). He concludes, "if we do not make this concerted effort, *I fear that sociology in the year 2000 will be no more advanced than it is today, though perhaps it will contain more specializations, theoretical schools, methodological cults and interest groups than even today, we can readily imagine.''*[1]

Rossi (1979) in establishing the theme of the 1980 American Sociological Association meeting, also recognizes the current chaotic state of the field. In his words: "The current diversity of paradigms, scholarly styles, and substantive concerns of the field of sociology is recognized in the theme, 'Chaos, Competition and Creativity' set for the 1980 ASA convention."[2]

Stryker's (1979) major theme in his presidential address to the annual meeting of the North Central Sociological Association in April was that: A common vision of the relevance of the full range of sociological concerns for the work of any sociologist is badly needed within sociology because the lack of clarity, coherence, and order in the identity *sociologist* is having detrimental effects on the profession. Stryker (p. 180) concluded: "Those who are most vulnerable in the professional system, the young and untenured, are most constrained to minimize risk in what they undertake. One might hope that, once some degree of security has been achieved, creative impulses will flourish. To believe that this will happen, in my estimation, is to underestimate the impact of the earlier socialization experience."[3]

At the 1979 annual meeting of the Midwestern Sociological Society, Joan Huber (1979) asked the panelists to address the question, "Where is the cutting edge in sociology?" Their comments portray images of the present state of sociology similar to those presented by Blalock, Rossi, and Stryker, although the panelists differed widely in their degree of optimism.

Morris Janowitz stated: "There is a strong feeling that the vast expansion in the number of sociologists and the corresponding increase in the number of academic journals have not been accompanied by a rise in scholarship standards or an enrichment of the sociological imagination. To the contrary, there is a growing recognition that we are producing a vast mass of mediocre research and analysis. Younger sociologists are concerned about the fragmentation of interests and the ceaseless growth of overspecialization which is almost forced or at least fostered by an increase in the number of highly specialized sociological publications. Is sociology condemned to a series of passing fads, without cumulative accomplishments? Can sociologists through self-critical discussion identify the foci of creative efforts? . . . From my point of view, the answer is clear. Sociology has no particular cutting edge or edges. Good and enduring work can be found scattered throughout the various subject matter interests and differing perspectives—[but]—one must look carefully and read widely to encounter that which is worthwhile."[4]

Norbert Wiley (1979) summarized his evaluation: "I see three big trends: a quieting down of theory, a growth of methodology to the point where it is nearly muscle-bound and a digging into epistemological questions." If we read him correctly, Wiley sees sociology as becoming atheoretical, overly methodological, and permeated with doubts about previously assumed knowledge and theory.

Robert Leik (1979), himself a mathematically inclined methodologist, sees the positive potential of many recent statistical developments for certain specific questions. However, he sees present methodologists as focusing almost exclusively on these developments and thereby ruling out the analysis of many important questions. In his words, "If there is only one approach to theory, in the sense of the kind and shape of acceptable relationships (asymmetrical, linear), then we do not even ask many potential useful questions. . . . In short, the only questions asked are whether more variables should be added, and in what order they should appear. There is not much creativity in such an enterprise" (p. 602). Leik also states, "As Marie Osmond has urged in a recent issue of the *American Sociologist,* we need more basic models to aid our thinking in more simple but powerful ways to capture the essence of causal flow, of dynamic feedback, and so forth. The notion of models implied here is not necessarily mathematical or statistical. In fact, physical model analogs of the subject matter are often the best aids to thinking and communication. Watson's and Crick's double helix, for example, is a way of structuring mentally the way the living system is presumed to operate. It aids planning research, developing further hypotheses, interpreting results, and providing the 'thesis' against which the dialectic of intellectual endeavor can formulate an antithesis. Such a move in sociology is indeed a cutting edge" (p. 601).[5]

We would argue that the last sentence should read "would indeed be a cutting edge," for the example of the double helix is from biology and not sociology. Furthermore, we think the reason the example is from biology is clear: that there simply is no comparable example from any of the social sciences. It is also perhaps worth quoting Leik on one other point because, although it is all too obvious, it is also often left unsaid by sociologists: "It is hard to find references to sociology in the popular media which do not suggest we are pretty unintelligible. It is also hard to find legislators, public administrators, and other policy makers who believe that sociologists can answer any of their questions'' (p. 601). For a more detailed exposition on the topic the reader might want to look at Gibbs' (1979) article, ''The Elites Can Do Without Us.''

As sociologists we have focused on sociology. However, similar comments are readily found in the areas of economics, social psychology, and political science. In short, the critique of sociology presented in these comments also largely applies to the other social sciences.

It is noteworthy that the critiques of sociology and its sister disciplines uniformly present a picture of fragmented, atheoretical disciplines whose work is often methodologically unsound, although often ''sophisticated.'' Except for Blalock, none of them points to a concrete direction in which we should proceed. And Blalock, crudely translated, appears to be saying, ''Do what we have been doing, but do it better and do it collectively.''

The Antibiological Bias

Nonetheless, leading sociologists have almost uniformly avoided a linkage with biology. As Petryszak (1979, p. 300) states (providing thirteen relevant citations since 1972): ''The belief maintained by most North American social scientists on the plasticity of human nature and its consequent dependence on culture has proved to be a rigid bulwark against the introduction of theories which maintain a biosociological viewpoint.''

As a separate part of his analysis Petryszak (p. 292) performed a content analysis of core themes in twenty-four introductory sociology textbooks. He found they presented the following basic ''assumptions about human nature, culture and the processes of social determination'' (the ordering follows the frequency of the assertion):

1. That any consideration of biological factors believed to be innate to the human species is completely irrelevant in understanding the nature of human behavior and society (eleven books).
2. That all differences between cultures are attributable to sociological factors (four books).
3. That human culture is comprised solely of the ideational and technological aspects of society, excluding any consideration of a biological basis (three books).
4. That culture is intimately linked with the process of social interaction, and that social interaction is the primary factor responsible for the development of the social self (three books).
5. That man's ability to learn and to be susceptible to the process of socialization and the opinion of others is due solely to the absence of all instincts in man (three books).

6. That culture is based on man's ability to learn and communicate through language (two books).
7. That instincts are asocial while culture is conceived of as being completely social in nature (one book).

Petryszak also found such statements as that by Bell and Serjamaki (1964, p. 22): "Wherever the theory [of] inheritance of human behavior exists there is also the possibility of the emergence of fascism."

Biological Undercurrents in Sociology

Although most social scientists, particularly sociologists, have been antagonistic to biological explanations of human behavior, there has long been a minority who have pushed for a joining of the social and the biological. As Petryszak (p. 291) notes, "In sociology today, a consensus exists that the human self is exclusively socially determined. Many sociologists maintain that any considerations of the biological basis of human behavior are irrelevant. The arguments which are made today in favor of the social determination of the human personality use the theoretical insights of the Pragmatists and Symbolic Interactionists as their source of reference and proof." Then Petryszak shows that these two interrelated perspectives are deeply immersed in assumptions regarding the biological determinants of the human self. In his words, "the Pragmatist and Symbolic Interactionist theories which are used presently to justify the rejection of all biological considerations of human behavior are themselves characterized by the use of assumptions about the biological basis of human nature" (p. 300–301).

Petryszak (p. 300) explains the rationale of the Pragmatist and Symbolic Interactionists' use of biology in the following manner:

> There are several different reasons why theorists such as Peirce, James, Dewey, Cooley, Mead, Thomas, Blumer and Mills have admitted to various biological assumptions about human nature. The major reason is that while sociologists were able to generate a theory of social determinism and social integration, they were nevertheless unable, using the same social deterministic assumptions, to develop a theory explaining and legitimizing individuality. It was their strong belief in the liberal ideals of rights of the autonomous individual which required them to give recognition to the significance of the individual actor in their own deterministic paradigms. While the social deterministic assumptions of Pragmatism and Symbolic Interactionism rationalized the need for social integration, the biological assumptions about human nature assured them that man was to some degree aloof and autonomous from society as a self-motivated actor within the social process.[6]

In an influential article, "The Oversocialized Conception of Man in Modern Society," Wrong (1961) points out that most sociologists have a very incomplete view of man. He discusses forces in man that are resistant to socialization. In his words, "It is not my purpose to explore the nature of these forces or to suggest how we ought best conceive of them as sociologists—that would be a most ambitious undertaking. A few remarks will have to suffice, I think we must start with the recognition that *in the beginning there is the body.* As soon as the body is mentioned the spectre of 'biological determinism' raises its head and sociologists draw back in fright" (p. 130). In discussion of biological determinism Wrong draws heavily on Freud. He notes that for Freud humans have a vague but powerful and impulsive nature and emphasizes that the proper social object of this nature is not selected in advance. According to Wrong (p. 131), "The drives or 'instincts' of psychoanalysis, far from being fixed dispositions to behave in a particular way, are utterly subject to social channeling and transformation and could not even reveal themselves without social molding." Thus for Wrong and for psychoanalysts in general, a human's social nature is profoundly reflected in his biological nature but he is indeed a social animal. Wrong goes on the emphasize that the Freudian and sociological conceptions of humans are nonetheless very different. "To Freud, man is a social animal without being entirely a *socialized* animal. His very social nature is the source of conflicts and antagonism that creates resistance to socialization by the norms of any of the societies which have existed in the course of human history" (p. 131). Wrong points out that socialization can mean two things. For Freud (and Wrong) it means the transmission to an individual of the particular culture of the society that individual enters at birth, as well as the acquisition of uniquely human attributes from interaction with others. However, Wrong argues, and we think correctly, that for most sociologists persons are "completely molded by the particular norms and values of their culture" (p. 132).[7] This in Wrong's view is the oversocialized conception of man.

The undercurrent of a biological explanation of human behavior started to emerge as a more prominent, if a fringe, perspective in the early 1970s (for example, Mazur and Robertson 1972, Mazur 1973, Van den Berghe 1974). However, with the publication of Edward O. Wilson's (1975) book, *Sociobiology: The New Synthesis,* the social sciences were confronted with an extensive treatise that had a very solid biological grounding in the studies of species other than *Homo sapiens;* it received wide popular attention, and it proposed a biological perspective on human behavior that was in direct opposition to the perspective held by mainstream social scientists. (The reader should be aware that in biology in many respects the present era is a "golden age" and this inevitably promotes a context in which biological explanation of behavior, however speculative, is apt to produce a favorable reaction in many knowledgeable individuals.)

Sociobiology asks grand questions, as can be seen from Wilson's idea of its scope. On page 4 of *Sociobiology* he says it is the "systematic study of the biological basis of all social behavior." *All*, he says. Thus he presumes "it explains all social behavior of all social animals." Or even further: "if one assumes evolution from a common archaic biological origin, then, biology explains all behavior of all social animals of all times, theoretically at least." That is an enormous compass. And there are consequences to positing that explanatory reach. First, in this view biology is obviously the queen of the social sciences, and the present sociologies and psychologies are merely (even if now large) subfields of concern. This, we cannot help but notice, has more than just analytical consequences. It is an assertion in a political domain as well, for there is a sense in which it represents a more severe criticism of the assumptions of liberalism than does current behavioral psychology or sociology. Because biological determinants are sought in the hierarchy of priority of sciences, biology is logically more basic than the sciences of the *symbolic social*. The notion is this: Man is an animal and only one of many who are social. So the study of the social, even the symbolic social, of man is properly the study of the natural. Thus, the inquiry into man's behavior is simply a subpart of the larger inquiry that concerns the behavioral sciences.

Seldom has a work generated as strong a response as Wilson's *Sociobiology*. In a review in *Scientific American,* John Bonner (1975, p. 131) writes: "Rarely has the world been offered such a splendid stepping-stone to the exciting future of a new science." Pierre L. Van den Berghe, reviewing the book for the sociological community, writes: "Having systematically ignored, indeed, rejected biological thinking for a half-century, sociology is now facing what is probably the most fundamental intellectual challenge of its history from the life sciences" (1976, p. 731). In this, the golden age of biology, "Books like Wilson's will make further resistance to biological concepts not only difficult, but patently silly" (1976, p. 731). The response was not limited to academia. For example, the front cover of the August 1, 1977 issue of *Time* magazine carried the cover story, "Why You Do What You Do: Sociobiology: A New Theory of Behavior," and it depicts a man and a woman as puppets. Although the popular reaction and the academic reaction to Wilson's work have been largely negative (for example, Petryszak 1979, p. 300),[8] these reactions have been counterbalanced by the development of a number of proponents of sociobiology in sociology as well as in biology. For example, among sociologists there are a rapidly expanding number of well-attended sessions on sociobiology at regional and national meetings, the *Biosociology Newsletter* is gaining a wide circulation (Baldwin and Baldwin 1980), and biosociology has become a recognized area of specialization regularly listed in the Graduate Guide to Graduate Departments of Sociology (1980).

Upon reading Wilson's (1975) *Sociobiology* one wonders what the attention and concern is all about. Most of the book is devoted to a very wide range of species other than human and is properly viewed as an excellent treatise on evolutionary theory. Since the modern synthesis of evolutionary theory was developed early in the century, neo-Darwinist thought has been the organizing principle of biology (Mayr 1978). *Sociobiology's* theoretical and empirical base is almost exclusively that of neo-Darwin evolutionary theory and research. As such, sociobiologists focus on *distal* causes—which they term the "ultimate causes"—of behavior. As Baldwin and Baldwin (1980, p. 10) note: "By claiming that natural selection provides the ultimate explanation of behavior, sociobiologists easily convince themselves that proximate causes need not be studied as autonomous influences on behavior." The powerful forces of natural selection are presumed to program the proximal mechanisms so that everything important is taken care of by genetic control. The proximal "machinery carries out the commands of the genes" (Wilson 1975, p. 23). Even the brain, the organ of thought and learning "is an extension of the gonads" (Ghiselin 1973, p. 968) geared "to favor the maximum transmission of the controlling genes" (Wilson 1975, p. 4).

Yet, when one turns to his final chapter, where he focuses on humans, one sees clearly that Wilson is sociologically and psychologically naive. One searches in vain for any discussion of social psychology or sociology. There is virtually nothing to indicate an awareness of what is contained in these or related fields. There is virtually no discussion of research or theory in these areas and one looks in vain even for citations of scholars in these areas. Two years later it is true that Wilson (1977, p. 136–137) does seriously address one theoretical formulation by one sociologist. That sociologist was Durkheim, the work was published in 1895; it was clearly polemical in intent and Wilson treats it entirely out of context. Perhaps even more damning is that in terms of neurobiology it was clear in 1975, and particularly 1978, that Wilson was and remains extraordinarily naive. Wilson appears to have no awareness of the research on the cortex and the implications this research has for understanding the determinants of human behavior. What Wilson (1975, 1978) does do is apply the basic principles of modern evolutionary theory to humans and then, using these principles, undertakes a number of speculative excursions. As this is the sociology in Wilson's sociobiology, one wonders why it is taken so seriously, particularly as the direct application of the evolutionary principle is very straightforward and neither controversial nor particularly innovative.

Given our negative evaluation of what has become known as sociobiology one might ask why it has clearly become a school of thought that cuts across disciplines. We think the reasons are obvious. First, there is wide agreement among social scientists that their fields are fragmented, that they lack a clear focus and direction, and that without a new focus their fields will stagnate. Second, it seems obvious to a number of social scientists that the biological aspects of humans play a prime role in determining human behavior. Third, in a world of turmoil, where the basic foundations of western civilization are clearly at question and we face the possibility of mass annihilation, some may find it tempting to tie human behavior to something as stable as biology. If we are correct on these points, then the idea of combining the social with the biological will clearly be attractive to a number of concerned academics. Furthermore, because social scientists rarely know much biology they are not in a position to evaluate the biological aspects of sociobiology. Also, because social scientists who are critical of sociobiology are also invariably blatantly ignorant about basic biology and their social deterministic ideology is very clear, it is not surprising that biologists feel they can readily dismiss most of the criticism by social scientists.

Beyond Wilson's (1975) *Sociobiology*

There have been two sets of works in sociobiology that attempt to go beyond Wilson's (1975) initial statement, both of which focus on humans. One of these is the series of works by Pierre Van den Berghe (1973, 1974, 1975), a sociologist who was writing at the same time Wilson's initial work was being produced. The second is another book by Wilson (1978), *On Human Nature,* which received the Pulitzer Prize.

Van den Berghe's (1973) first book, *Age and Sex in Human Societies: A Biosocial Perspective,* applies an informed anthropological and sociological perspective to uniformities in age and sex roles in human societies. By looking at uniformities in sex- and age-related patterns of social behavior that occur in a wide range of societies, Van den Berghe argues that in all likelihood the consistent patterns reflect innate sex and age differences in capabilities. In his article in the *American Sociological Review,* Van den Berghe (1974) suggests that "Homo sapiens rate high on territoriality, hierarchy and aggression and that these forms of behavior are biologically predisposed" (p. 777). Again his work reflects considerable anthropological knowledge. However, his use of biological information is selective. For example, his argument that *Homo sapiens* have a high level of hierarchy compared to other mammals ignores the fact that the high level of hierarchy is made possible by technology and the comparison with certain mammals who, for reason of food alone, need to be widely dispersed would seem inappropriate. He also simply ignores the qualitative differences in intellectual capacities and the ability to utilize symbols that clearly differentiate *Homo sapiens* from other species.[9]

Van den Berghe's (1975) major statement, *Man in Society: A Biosocial*

View, starts out with the following assertion: "By and large, the more significant the question is, the less capable sociology is of giving an answer. And that inability has a great deal to do with the partially indeterminate, or *selfdetermined* character of human behavior" (p. 7). He goes on to criticize what he sees as the ethnocentric, ignorant, and overly socially deterministic view of sociology and to condemn "the failure of sociologists seriously to consider the biological bases of human behavior and the complex interplay of organism and environment" (p. 26). To a large extent his book reads as an anthropological introductory text with a strong biological focus. Although the biological emphasis would be uncharacteristic of most introductory anthropology texts, the importance of biology emerges only on the basis of inference and the anthropology used is sound.

Unfortunately Van den Berghe's book is really not informative about interaction between biological and social phenomena. This is evident in his citations, which are primarily anthropological, and secondarily sociological. The few references to biological works tend to be to those widely recognized as popularized syntheses (for example, Adrey 1961, 1966; Lorenz 1966; Morris 1967, 1969; Tiger and Fox 1971). Van den Berghe's (1974, 1975) focus on the unusually aggressive nature of man is clearly distorted, as a reading of Wilson (1975) on nonhuman animals would show. Similarly, his denial of a genetic component to differentiate in intelligence (for example, Van den Berghe, 1973, pp. 36–38) clearly is inconsistent with a well-established (if disliked) body of evidence (see the chapter by Taubman). In summary, Van den Berghe's work, although it has been extremely controversial among sociologists (Larson 1976), when carefully scrutinized simply points to uniformities in human behavior that are suggestive of biological determinants, and it reflects a lack of biological knowledge.

Wilson's (1978) book *On Human Nature* is a very serious effort to apply sociobiology to humans. As Wilson states, this book is "simply the extension of population biology and evolutionary theory to social organization" (p. x) and, further, that it is an "uncompromising application of evolutionary theory to all aspects of human existence" (p. x). As he notes, his book poses "two great spiritual dilemmas" (p. 2). "The first is that no species, ours included, possesses a purpose beyond the imperatives created by its genetic history" (p. 2). The essence of his argument is "that the brain exists because it promotes the survival and multiplication of the genes that direct its assembly. The human mind is a device for survival and reproduction, and reason is just one of its various techniques" (p. 2). For Wilson "sociobiology can account for the very origin of mythology by the principle of natural selection acting on the genetically material structure of the human brain" (p. 192).

Wilson's second spiritual dilemma is that "innate sensors and motivators exist in the brain that deeply and unconsciously affect our ethical premises; from these roots, morality evolved as an instinct" (p. 5). Wilson goes on to indicate that science "may soon be in a position to investigate the very origin and meaning of human values, from which all ethical pronouncements and much of political practice flow" (p. 5). In short, Wilson thinks biology, particularly the evolutionary theory of sociobiology, will undercut the "religious" mythologies and feelings of ethics and morality, which he views as adaptive mechanisms.

Wilson recognizes we are an evolving species that has evolved to the point where (1) many of the adaptive psychological and social mechanisms are no longer effective and (2) we will soon reach the point where we can manipulate our censors and motivators on an evolutionary scale. In his words, we will soon confront the question of "which of the censors and motivators should be obeyed and which ones might be curtailed or sublimated." Wilson offers a solution to these dilemmas, namely, to put our faith in science, particularly biology and most particularly sociobiology. In his words "to chart our destiny means that we must shift from automatic control based on our biological properties to precise steering based on biological knowledge" (p. 6). But, what if "we" (that is, sociobiologists) make the wrong choices? Yet even about this issue, according to Wilson, we probably should not worry, for as he (1975, p. 208) says "perhaps there is something already present in our nature that will prevent us from ever making [the types of] changes . . . [that affect] . . . the very essence of humanity" (p. 208).

In summary, Wilson has developed a grand scheme. But this scheme is based largely on speculation, is linked almost exclusively to one aspect of biology, and reflects a vast lack of comprehension of the psychological and the social. Perhaps more importantly it lacks a recognition that the biological, the psychological, and the social all routinely function as variables causally affecting each other, and that the direction is only sometimes from the biological to the psychological and the social.

Outline of This Book

It is our purpose in the present book to bring together works that reflect some of what is known about biopsychosocial interaction. Because we perceive our audience as largely composed of social scientists, this book tends to emphasize the importance of biological processes. (Regarding this emphasis we would note that if we perceived our audience as primarily composed of persons with a biological perspective we would have placed more emphasis on how psychological and social processes affect biological processes.) Our emphasis, however, is only a matter of degree and we think scholars with either a biological or social science perspective would benefit from an acquaintance with the materials in this book.

Chapter 2, by Carpenter and Gove, states the epistemological issues involved in trying to study human behavior. Marcus Feldman and L.L. Cavalli-Sforza are two eminent evolutionary theorists who provide a theoretical critique of Wilson's theory of evolution and an alternative model in chapter 3. Their chapter is followed by one by G.C. Davis, M.S. Buchsbaum, and W.E. Bunney, Jr. that demonstrates how psychological factors are related to the production of biological processes affecting pain. In chapter 5, Herbert Weiner demonstrates the complexities of the interaction of biological, psychological, and social processes in the development of illness. Ingrid Waldron (chapter 6) explores the social, psychological, and biological causes of sex differences in morbidity and mortality. C. Dawn Delozier and Eric Engel (chapter 7) take sexual differentiation as a model for the interaction of genetic and environmental factors in physical and psychological development. Jacquelynne Eccles Parsons discusses the biological and psychological factors influencing sex role behavior.

The book then shifts focus to look at determinants of social success and deviance. In chapter 9, Paul Taubman assesses the extent to which environmental and biological factors are related to earned income. Juan B. Cortes looks at the interaction of biological, psychological, and social factors in the development of delinquency in chapter 10. Chapter 11, by Kenneth K. Kidd, discusses how genetic and social factors are related to behavioral disorders. Chapter 12, by Gerald L. Klerman and Gail Schlecter, demonstrates how psychopharmacology, along with changing social policy, has totally restructured (1) the life careers of persons who become mentally ill, and (2) the way in which the mentally ill are socially processed. Chapter 13, by Linnda R. Caporael, which presents a largely biological explanation for the occurrence of the Salem witch trials, is presented as an argument in direct conflict with Kai Erickson's influential sociological explanation of the Salem witch trials presented in *Wayward Puritans*. G. Russell Carpenter presents in chapter 14 a discussion of the complex ethical, scientific, and pragmatic issues involved in the social regulation of biological processes. Finally there is a brief summation by Gove and Carpenter of the general themes of the previous chapters, a discussion of needs and direction of future inquiry, and a brief example of how social, psychological, and biological processes can be conceptually combined in a manner that may lead to a more fruitful approach to the study of human behavior.

Notes

1. Hubert M. Blalock, Jr., "The Presidential Address: Measurement and Conceptualization Problems: The Major Obstacle to Integrating

Theory and Research," *American Sociological Review* 44 (December 1979): 881–894, emphasis added. Reprinted with permission. Blalock details nine specific problems and discusses their solutions. In essence he is asking for a concerted methodological effort toward measurement and causal modeling to enable sociologists to deal with much more complex areas of variables and analysis than are now within their grasp. Although we largely agree with his diagnosis of the state of the field, it should be clear from the present book and from Gove and Carpenter (1980) that in our view his proposed solution points in precisely the wrong direction.

2. Although Rossi's diagnosis of the current state of the field of sociology is similar to Blalock's, contrary to Blalock (1979), Rossi is optimistic about the future of sociology. In his words, "This diversity, I believe, is a cause for celebration. We are going through a period of great creativity. What appears to be chaos is, in fact, competition among varying views that will lead in the end to the emergence of a field that will have a sense of where it is going, a conviction that it is an important intellectual endeavor, and conviction that is has something to say to a society that needs its special point of view." In our view Rossi's statement is based on faith and hope and little else. (Peter H. Rossi, "Rossi Recognizes Diversity in 1980 Annual Meeting Theme," *Footnotes* 7, no. 6 (August 1979):1. Reprinted with permission.)

3. Sheldon Stryker, "The Profession: Comments from an Interactionist's Perspectice," *Sociological Focus* 12 (August 1979): 175–186. Reprinted with permission.

4. Morris Janowitz, "Comment: Where Is the Cutting Edge of Sociology?" *Sociological Quarterly* (Autumn 1979):591–592. Reprinted with permission.

5. Robert K. Leik, "Comment: Where Is the Cutting Edge of Sociology?" *Sociological Quarterly* 20 (Autumn 1979):601–602. Reprinted with permission.

6. Nicholas Petryszak, "The Biosociology of the Social Self," *Sociological Quarterly* 20 (Spring 1979):291–301. Reprinted with permission.

7. Dennis H. Wrong, "The Oversocialized Conception of Man in Modern Society," *American Sociological Review* 26 (April 1961):130–132. Reprinted with permission.

8. It even resulted in an organization of academics aimed specifically at undoing the harm produced by the work of Wilson and his collaborators. (*See* Sociobiology Study Group of Science for the People in 1976).

9. For example, Van den Berghe (1974, p. 781) states, "No mammal species comes close to homo sapiens in the elaborateness and multiplicity of invidious distinction" as an illustration of man's tendency for dominance and hierarchy, and he never considers that this multiplicity of distinction may simply reflect a difference in the intellectual capacity to utilize symbols

and not necessarily a rigid hierarchical structure. In fact, it is possible to argue that *Homo sapiens'* ability to make use of a "multiplicity of invidious distinctions" can, in certain situations, function to weaken the degree of dominance and rigidity in a hierarchy.

References

Adrey, Robert. 1961. *African Genesis.* New York: Atheneum.
———. 1966. *The Territorial Imperative.* New York: Atheneum.
American Sociological Association. 1980. *Graduate Guide to Graduate Departments of Sociology.*
Baldwin, John, and Janice Baldwin. 1980. "Sociobiology a Balanced Biosocial Theory?" *Pacific Sociological Review* 23 (January):3–27.
Bell, Earl, and J. Serjamaki. 1964. *Social Foundation for Human Behavior.* New York: Harper and Row.
Blalock, Hubert. 1979. "The Presidential Address: Measurement and Conceptualization Problems: The Major Obstacle to Integrating Theory and Research." *American Sociological Review* 44 (December):881–894.
Bonner, John. 1975. "A New Synthesis of the Principles That Underlie All Animal Societies." *Scientific American* 233 (October):129–131.
Dawkins, Richard. 1975. *The Selfish Gene.* New York: Oxford University Press.
Ghiselin, M. 1973. "A Radical Solution to the Species Problem." *Systematic Zoology* 23:536–544.
Gibbs, Jack. 1979. "The Elites Can Do Without Us." *American Sociologist* 14:79–85.
Gove, Walter, and G. Russell Carpenter. 1980. "The Validity of Present Paradigms in the Social Sciences." Meeting of the Eastern Sociological Association, Boston, March.
Huber, Joan. 1979. "Comment: Where Is the Cutting Edge of Sociology?" *Sociological Quarterly* 20 (Autumn):591–603.
Janowitz, Morris. 1979. "Comment: Where Is the Cutting Edge of Sociology?" *Sociological Quarterly* 20 (Autumn):591–595.
Larson, Roger. 1976. "Review of Men in Society: A Biosocial View by Pierre Van den Berghe. " *Contemporary Sociology* 5 (January):5–9.
Leik, Robert. 1979. "Comment: Where Is the Cutting Edge of Sociology?" *Sociological Quarterly* 20 (Autumn):600–603.
Lorenz, Konrad. 1966. *On Aggression.* New York: Harcourt, Brace and World.
Mayr, E. 1978. "Evolution." *Scientific American* 239(3):47–55.
Mazur, A. 1973. "A Cross-Species Comparison of Status in Small Established Groups." *American Sociological Review* (38):513–530.

Mazur, A., and L. Robertson. 1972. *Biology and Social Behavior.* New York: Free Press.

Morris, D. 1967. *The Naked Ape.* London: Jonathan Cope.

———. 1979. *The Human Zoo.* London: Jonathan Cope.

Petryszak, Nicholas. 1979. "The Biosociology of the Social Self." *Sociological Quarterly* 20 (Spring):291–303.

Rossi, Peter. 1979. "Rossi Recognizes Diversity in 1980 Annual Meeting Theme." *Footnotes* 7(August):1.

Stryker, Sheldon. 1979. "The Profession: Comments from an Interactionist's Perspective." *Sociological Focus* 12 (August):175–186.

Tiger, Lionel, and Robin Fox. 1971. *The Imperial Animal.* New York: Holt, Rinehart and Winston.

Van den Berghe, Pierre. 1973. *Age and Sex in Human Societies: A Biological Perspective.* Belmont, Calif.: Wadsworth.

———. 1974. "Bring the Beasts Back In." *American Sociological Review* 39(6):777–788.

———. 1975. *Man in Society: A Biosocial View.* New York: Elsevier-North Holland.

———. 1976. "Review of Sociobiology: The New Synthesis." *Contemporary Sociology* 5 (6):731–733.

Wiley, Norbert. 1979. "Comment: Where Is the Cutting Edge of Sociology?" *Sociological Quarterly* 20 (Autumn):593–600.

Wilson, Edward. 1975. *Sociobiology: The New Synthesis.* Cambridge, Mass.: Harvard University Press.

———. 1977. "Biology and the Social Sciences." *Daedalus* 106 (4):127–140.

———. 1978. *On Human Nature.* San Francisco: Jossey-Bass.

Wrong, Dennis. 1961. "The Oversocialized Conception of Man in Modern Society." *American Sociological Review* 26 (April):183–193.

2 The Study of Human Behavior: Some Epistemological Questions

G. Russell Carpenter
and *Walter R. Gove*

There are some terms we use in science as a necessary part of the enterprise, terms that do the work we require of them because they both summarize too much and specify too little. *Civilization* is one of these terms, and *nature* another. Each has a grand status in science and yet, neither of them can be specifically defined. Yet we use them, and they work for us because they form a kind of theoretical or contextual background for our more specific concepts, issues, and concerns.[1] There is not, for instance, a paradigm that stipulates, completely and without residual, what either of these terms finally means. Definitions of them are more apt than not lengthy narratives, because otherwise there always seems to be something left over, or something more that still needs to be explained or further stipulated. Both terms, in fact, seem to be best described in what Gilbert Rile (1949) terms "thick description," because they are concepts that are capable of being described neither in the elegance of mathematical equations nor in the astringency of a set of logical propositions. But they are, nevertheless, absolutely essential to the work we do.

Theory

As important as they are, it appears that these two terms will work for us only so long as they are used separately or are kept apart in our discussions.[2] Which is perhaps no more than saying that there are cultural determinists on the one hand, and biological determinists on the other, and never the twain should meet. Or at the minimum it points out a fundamental division in modern thought: it indicates that there have been two basic approaches to explaining human behavior.

The division here is an old one and as such it travels under a variety of named pairs. But regardless of what the pairs are called the same cleavages and antinomies exist between the concepts of nature and nurture, instinct and thought, reason and desire as exist between the more classic division of mind and body. The basic division may go back as far as the Greeks, but certainly it is as old as the work of Descartes. And although there are clear differences between the things each of those pairs describes, as well as differences between our thought and the Enlightenment's images of things, this division has remained fundamental to our explanations of behavior. And, it has been especially fundamental (as well as troubling) to our explanations of human behavior. Because man can be seen from the perspective of either side of the division, but then only so long as the presence of the other side *is tacitly assumed at the same time* in the discussion.

We are, for instance, in the words of the theory of evolution, a biological being, a species, and like other species a product of a long history of chance, luck, error, and selection. Which is to say, we are animals, and as an evolutionary product we are still evolving. In short, we are an example of both nature and the processes of nature just as are the horse and the tree. Since the time of Darwin no person familiar with this basic assumption in Western thought would deny this. The belief that man is the product of evolution is a characteristic of our thought that is both fundamental and *foundational* to it. At the minimum, even to the most culturally oriented social scientist, our biological basis provides the limits of our behavior, as it is assumed to be the foundation upon which life and all of its conditions, including culture, are built: without biology no society.

But still, even in that fundamental a recognition of our basic biological nature there is slippage across the divide between civilization and nature, because while evolution may seem as natural, something we share with the rest, its theory is clearly the product of reasoning, and so is on the other side of the divide. The problem lies in the fact that Darwin included himself within his own theory, that is, saw himself as evolving animal and yet could not explain himself as theorist by the theory. His position, or more technically the position of his theory, is left ambiguous. The question that can be put to him is simply this: Is his theorizing about evolution an act of survival, or is it something else? And, if it is something other than a survival strategy, then, what else is it? The classical tendency was, of course, to solve this dilemma of the position of man and his works in nature's schemes by giving nature two distinctly different forms: nature and human nature. And while classical thinkers recognized that man was like the other animals in that he ate, and breathed, and died, it was also thought that he had a basic nature different from the rest of the animals. And for them if this different nature was interwoven with the rest of nature, for classical thought this was because human nature could turn back toward nature and by naming and arranging things, know the rest. So human nature was the last link in the great chain of being. It was assumed that the development of symbolic and rational thought made all the difference and created a "human nature" different from the rest.

Yet, even with that separation and that difference, *Man,* this partially sovereign, complex, empirical object that forms so many of the themes of our modern concern, makes no appearance in classical thought (Foucault 1970, pp.303–312). The modern image we hold of ourselves as man, this being whose nature works in accordance with the laws of biology, economics, language, culture, and society, and who exists always at the meeting point of all those, was simply excluded from view in the classical opposition of nature with human nature. Darwin's dilemma simply was not perceived as a dilemma in classical thought. The concept of *human nature* worked for them as *the foundation* for knowledge, it summarized everyone's position regarding the understanding of things, it was the birthright belonging to all born into the species.

But things are much different for us, and in our thought. The relationship between nature and civilization is not only a dilemma in our work, and one that is creating a crisis in our conceptualization of man and society, but more important, it is related to a growing crisis in our civilization. In the words of ecologically informed political theory, civilization cannot be a declaration of independence from the natural world or from the bonds of our biology. There is no such exemption. Indeed, man's biological niche is now a topic in our political discussions. There has been a growing recognition that the acceleration of environmental damage throughout the world has been largely the consequence of the scientific enterprise, and in light of that, we cannot continue to maintain the analytical separation between an economy of the mind and the economy of the nature.[3] For in our era social power is situated and articulated at the level of the population, and this growth and change in the development of technique has direct bearing on how man will survive (if man survives) as well as on the whole ecology of the planet.

Modern industrial society integrates the production of life so completely that the mechanisms of modern power must work more to generate, foster, order, and control the population than they must work to delimit, inhibit, or exclude.[4] An end point of this process is surely genetic manipulation by means of scientific and medical technique. So while biology is the foundation of our civilization, as it is of any other, in our era civilization has also developed the ability to directly manipulate our biology. And this ability, which we have recently acquired, has further blurred the line between the natural and the artificial.[5]

Yet, the social science models, especially those that depend upon the hierarchy of sciences that Auguste Comte delineated, create their explanations by maintaining a division between biology and society; these models no longer describe the lived reality, and this insufficiency leaves us with two intellectual tendencies in evidence in the literature.

The first, and by far the more prominent of the two, is the tendency to ignore the grand issues and macroconnections by focusing research on a particular issue that can be easily specified with a very few hypotheses.[6] This approach seems to be low-risk science because of its emphasis on methodological sophistication and mathematical formalization. The concern in this approach is mainly with obtaining precision and rigor, which is the first order of methodological business in science. This taken by itself is a laudable goal; but this specialization, when viewed in terms of the growth of knowledge, does not seem to produce its intention, which we assume to be the full articulation of the issues involved in a paradigm of science. Instead, it is getting increasingly hard to say just what is the domain of a discipline. We are left with roughly defined departments. And, the further we go along here the more all of us seem to be writing individual discourses.[7]

In this intellectual climate, the master concepts that set the discipline's boundaries, such as the concepts of society, culture, personality, mind, or economy, tend to move into the intellectual background. And from that position what use they have is metaphoric. The result is a theoretical crisis of the first magnitude in the study of human behavior. We have come to a place in our disciplines where there is much empirical work that shows little or no evidence of theory. Thus, one is left wondering if this low-risk science is science at all, and whether there is any real meaningful distinction among our disciplines. But more importantly, others are left questioning the worth of all of it, and some of those others finance our work.

The other tendency in the literature on human behavior is the attempt to make the grand connections, and often with the use of metaphors. The return of Marxist analysis to recent American thought is surely this. The temptation in Marxist analysis is to connect the pieces of lived reality by the metaphors of oppression and the ubiquity of production. Everything relates to everything else because the metaphors are so wide, or so loose that everything can be related. But then the details are always the problem. Marx, for example, never quite explained how it was that class membership actually constrained or warped the consciousness of those within. His discussion of class was left unfinished, and that is how it remains today.

So the metaphors that connect the macro with the micro *work because they indicate but do not specify the actual connections.* Because of this the criticism of metaphoric connections is always that such work is unscientific. And when you think about this, it is a criticism generally leveled at Marx, and Freud, and most recently, at the work of Edward Wilson (see Caplan 1978). Freud's concepts of id, ego, and superego are surely metaphors, and not themselves organizations to be found in any particular brain, which means they cannot be empirically refuted. And Wilson's use of the concept of human nature is much the same. Wilson uses it as a metaphor; he calls his book titled *On Human Nature* (1978) "a speculative essay," and that it is. In fact, the concept of human nature works for him because it summarizes so much and specifies too little. Using it, he connects man's biological or genetic foundation to man's social behavior. But, if man's biological endowment is billions of brain cells, which together "program" us in certain ways, his social behavior is almost as complex, if, indeed, not more so. In fact, man's behavior is so multifaceted and so tied to a variety of symbolic connotations that its accurate description continues to escape our analytical abilities (which are, of course, themselves social behavior).

The concept of human nature makes an appearance again because, as with classical thought, it appears to pull us out of the morass of trivia. But the problem with its current use is that the concept of man has changed. We conceive of man as if he has a singular nature, but we conceive of him as expressing this nature within the systems of language, culture, economics, and history (Foucault 1970, pp. 312–318). If human nature works its way into behavior, we say it does so in the terms of those systems. The assumption is that man simply does not stop being psychological because he is biological, nor does he stop being social, or economic. Instead our models see him as all of those at once. And even for Wilson he is sometimes more of the latter than the former. Wilson does not see human nature as knowing human nature, instead he sees sociobiology as doing that.[8] And, in that difference, we again find Darwin's dilemma. If sociobiology is not genetically determined but is something else, then what else is it? If it is reason, and solely the produce of a "high" civilization, and not in biology's sovereign domain, then is this not the same kingdom that has come to manipulate genetic endowment by rational technique? So sovereignty is still the issue: Is reason king, or is biology?

Wilson, like Marx and Freud before him, attempts to make use of Darwin's insights in the study of human behavior (Wilson 1975, p. 4). But, like Darwin, whose theory was both in and out of its own domain; and like Marx, who claimed that ideology affected all social theory but his own; and like Freud, who argued that the unconscious affected consciousness but not significantly in the case of his theory; Wilson seems to stand in biological determinism and reason at the same time. But, like the rest, the mechanics of his exemption are never clearly worked out. The problem is, of course, all of the connecting metaphors these thinkers use both summarize too much and specify too little, yet without those metaphors each has little to say.

Issues

Regardless of how it is studied, human behavior always seems composed on the middle ground. If we argue that human behavior is the product of psychological factors, then we must admit that these psychological factors interface with two distinctly different domains. As we attempt to focus on the context in which the psyche functions, we immediately confront the contexts of society and history. All personalities exist in societies, and all societies exist in history.[9] No psychological determinist, regardless of his ardor in arguing for the psychological determination of behavior, would deny the presence of society or its important constraint upon human behavior. For Freud, civilization was constraint and the source of discontent and always there (Rieff 1961). In his model of the mind, *father* is a term almost completely replaceable by the concept of society, and in his theory there were no fatherless personalities, even if there were orphans. In contrast, if we focus our concern in the other direction, that is inward, we confront the interface of the psyche and the soma. At this point man becomes seen as a biological being with a species legacy that provides the biological foundations of our psyche. Thus to the degree that human behavior is argued to be primarily psychological that argument must assume that human behavior is cast upon a middle ground between our particular forms of society and the particulars of our biology.

If we put psychology aside, as sociologists are inclined to do, and argue that human behavior is socially determined, then we are left with the need to explain the experience of an implicit individual psychology, of which all of us are intuitively aware as we go about our daily activities.[10] Furthermore, even the most extreme social determinists will admit, when pushed, that there are always deviations around a norm, individual differences in style (that may be thought of as the expressions of personalities).[11] Once these are admitted, then even the cultural determinist is forced to admit that human behavior seems forged on the middle ground—in part psychological and in part social.

And even if we were to argue that we should put both of these parts to one side and assert that human behavior is biologically determined, then we are left with the demand to explain just how it is that biology would explain and answer that very argument: How it is that biology determines rationality—including the rationality in the biological determinist argument. But, then, no one would use a sociobiology to warrant or to justify a field of sociobiology. If sociobiology explains human behavior, it does not seem able to explain the behavior of writing sociobiology, even though, clearly, such lies within its topic domain. Even Wilson is forced to admit that "genes have given away most of their sovereignty." (1975, p. 575). Evidently then, sociobiology is seen by its adherents to rise in the silence of that absence of biological determinism.[12] This means that those making the sociobiology argument must see that argument as based much more on a philosophy of science than upon a biology of knowledge, while at the same time they must admit that both approaches to knowledge are possible, indeed, necessary to their argument. And so again, if we take sociobiology as the starting point we are forced to conclude that human behavior is constructed on the middle ground, being in part biological and in part psychological.

To the degree that economics demands an exegesis of value and a theory of desire along with its utilitarian calculus, then, economic behavior seems human and cast on the middle ground halfway between a social system of production and the articulation of human needs. And so too it is with history, anthropology, and the rest of human studies. No explanation seems to be totally adequate in any of these areas. Thus, perhaps, it is best to say that human behavior takes place *within* certain constraints, conditions, and determinants, some of which are biological, some of which are psychological, and some of which are social. Put another way, if human behavior is economic, it is so *within* an economic system; if it is biological or social it is so *within* the conditions of our biological or social inheritance. This is simply another way of saying that human behavior is always all of these at once (Foucault 1970, pp. 326–327). When humans behave psychologically they do not escape the economy, or their history, or the society, but instead they are psychological within those systems.

Modern thought sees behavior as always performed within constraints of which we are usually unaware, and these constraints *are a consequence of history.*[13] As biological beings, our behavior is limited by our biological foundation, which is itself a product of evolution, thus historical. Furthermore, it is likely that the evolution of the species has not reached its culmination. Or, if we focus on constraints of the economic system, then perhaps, Marx was correct and the forms of production and labor are historically determined and as such must still have transformations to undergo.[14] And so it is with the forms of society and the manners of personality integration: If ours is the age of narcissism, we know this, in part, because other ages were not; that is, we make a comparison of ours with theirs. Society and personality do not form a monotonic relation with each other when differences in society or social history are considered; and so it would appear there are still different journeys to take, different forms of psychosocial adaptations to be lived, and a new species to be lived with.[15] Put simply, our biological potentials and constraints, as well as our present culture, including that part of it we know as science, is located in a place in history that we now call the present. If man has a nature surely that nature changes, and if man's knowledge has a nature surely it changes also.

All of which indicates that concepts such as *human nature, civilization,* or *animality,* indeed, even the concept of *man,* are essentially contestable because we portray ourselves—every one of us—as thinking and reasoning (even about those concepts' definitions) from within their limits. It simply is no longer easy to point to something manifestly different than ourselves— some noncultural man for instance. There are no longer truly counter or negative instances to cite. And if we attempt to subtract civilization from what we are, we do not arrive at a foundational residue that is human nature or the forms of wildness in our nature. For unlike those in the Enlightenment, we no longer claim to know of the savage. Natural man is, for us, social man; we assume it to be the same game just in different circumstances. And if we claim to know civilization, we cannot claim to know it by stipulating what its absence is. All men live in society; there are no men without society. That is the modern assumption, there simply are no counter examples.

Thus mythical phenomena such as unbridled human nature are defined by assumption rather than by empirical observation. There simply is no case to observe. One cannot identify such phenomena by simply subtracting (or controlling) for the effects of other phenomena. Behavior comes whole, any division requires an assumption. Everywhere, regardless of the form society takes, at its core we assume it is the same: Men are always social. The differences that exist, the ones that make history more than sterile repetition, are still only differences bounded by the basic constraints of the social and the biological on man's schemes.

So what distinguishes our thought from classical conceptions is that for us, human nature is no longer conceived of as something distinct from the rest of nature and in opposition to the rest of nature. In our conceptions, human nature seems to emerge *from within* nature's conditions, as if nature allows us to think this, and to the degree that we find difference, it is not a consequence of opposition but of evolutionary processes that led us down a particular path while other species followed another. With the rest of nature we share a physical and a chemical constituency. With other animals we share an animal nature. However, the critical difference that sets us apart from other animals is that our biological nature coexists with a symbolic existence and thus we appear to ourselves (in symbols) as biologically symbolic beings that are fundamentally different from other animals.

It is in the symbols that man can turn back upon the foundation of his actual experience and, because of that gained distance that some call scientific objectivity, change the initial constraints of this existence, and even the constraints determining this thought. If the arrangement of DNA is the foundation of human civilization, then civilization has come to the place where it is possible to engineer the genetics determining civilization. At that point, chance and evolutionary necessity change their relation such that the historic biological foundation is no longer foundational in the same sense as before, and conscious reason can no longer be conceived as opposed to the unconscious mechanics of biology. It is in recognition of this point that all the classical paradigms of human nature, with their antinomies of conscious reason and unconscious desire, their separation of nature and nurture, their opposition of the cultural with the natural, collapse, and a new order of question emerges.

But then, what new questions? If the domain of human behavior's conditions stretches from the innate rules of species existence (biology) to the individuality of consciousness, then the new questions can be those that ask about the nature of the boundaries or the constituting conditions of behavior. Questions that concern the biochemical basis of the brain, and thus the constituting conditions of the mind's operation, have, at the moment, undefined boundaries. The search can begin there. In one form, these can be very specific questions that ask about the interaction of biochemical and psychological process. At this level, we can ask, for example, are the diseases of the brain deficiences of a structural nature that can be altered or reengineered in some way? Alternatively, we can ask, can these biological changes be produced by placebos that produce a biochemical change? This last question suggests that brain conditions are altered by systems of belief, just as warts can be removed by suggestion. These are obviously clinical questions, but they cannot be asked unless one combines a biological and a psychological focus in the concern.

There are theoretical questions here also. For instance, to what extent is the individuality of consciousness constrained by the limits of biological potential, or to what extent is it determined by the necessary social integration of a species? Or, what is the basis of altruism and what are its limits, given the individuality of consciousness? Or, is memory primarily a biological phenomenon and, if so, what would changing its biological (or individual) determinants do to a society? Would remembering the past differently change the nature of culture and thus the world in which we live and so the way we behave? Freud assumed that primary repression was a constant biological condition of the mind, but is it constant? Must it be? Freud simply assumed it was and concerned himself with secondary repression and the relationship of neurosis and the past (Freud 1949). But now we can return to that issue with a different analytic: How is memory biological and, since memory is necessary to culture, how is culture constrained by that biological basis? When asked this way the line blurs between that which is biological and that which is cultural; culture must be seen to exist in biological conditions because biology poses limits on the range of memory. The two cannot be separated.

And very different questions can be asked regarding the gene pool. Some find these questions awkward and are uncomfortable with them (see Caplan 1978). But in this, the golden age of the biological sciences, some of these questions are forced on us by the emergence of new technical possibilities enabling us to control our own biology. Now there are parents for organisms created in the laboratory, and in our era the gene pool has become an object of political thought and debate.[16] So in a sense, these questions, which some find so uncomfortable to ask, come to this one: In what ways, if any, is birth a political act? And this is not just an academic question. When a population can select the sex of its offspring, because this is an available biotechnique to them, will that population's sex ratio change, and if it does, whose concern is it, and what are the political and social consequences? This political question, involving as it does a right to privacy, and the most intimate of behaviors, has important public consequences. And this question, while only one of many, is typical of modern political questions concerning the limits of political control regarding biological engineering.

There is still another level of questions as well, concerning man's behavior and ecological scarcity. Human behavior exists not only in its biological and psychological conditions, but also in the conditions of society. But societies exist on an ecological base. As the current crises indicate we can no longer afford to think of an unlimited separation of society from nature. For while the study of man and society cannot be reduced to biology alone, at the same time one cannot ignore the fact we live in a biological world. Ecological scarcity demands that we deal with this issue, for there are rules of species existence that simply cannot be transcended.

Now clearly, of all living organisms, man is allowed the most open and flexible genetic program (Jacob 1973). Yet, surely there are important limits to that flexibility, and these limits can be sought. When behavior is viewed in terms of biological evolution, then questions concerning the limits of ability inevitably arise. The biological program in humans gives the organism an ability to learn, but it also imposes restrictions on what can be learned, when learning is to take place, and the conditions under which learning will occur (Piaget 1971). It is clear that there are limits on the amount of information that may be encoded, limits to the length of time that it will be stored in memory, and limits on the power of the mind to retrieve stored information. But are these limits absolute? Even if we change the biology? There are structural limits, we usually look for them in fullgrown adults, so other questions can revolve around development. These questions focus on time and on the issue of how human behavior evolves as the organism develops. These have been the focus of much psychology. But now they can be addressed with the precision allowed by the structural study of brain development. We can now ask, what biological changes occur with change in the socialization process? This question in turn involves another form of question: How does one system merge into another? More complexly put, how does a set of integrated systems—physical, chemical, biological—produce a psychological system on the basis of their integration?[17] And, should we want to go further and ask an even more sophisticated question, we can ask under what conditions can a level of emergence turn back and change the conditions of its own emergence. That is, we can focus not only on the emergence of society from biology—the traditional approach—but also on how biology comes to be shaped by society. For it is now clear that not only is biology a component of society but society also shapes biology—even if only in the way it produces food or illness.

Yet still another form of question can be asked concerning the interactions of the systems. With these questions, the concern for individual consciousness fades into the background and the attempt is to understand the interaction between social and biological systems. Much of the new naturalism focuses here. However, because the domain of human behavior involves the interaction of biological, psychological, and social phenomena, questions can be addressed at all these levels, and perhaps more important, at all the possible levels of interaction. Thus the domain is wide open; one need not be constrained by the simplistic assumptions of existent sociobiology.

In modern thought about man, the word *social,* like the word *behavior,* has two natures. On the one hand it is *the* social, and the word is used and intended as a noun that depicts some selected sets of human relations—including the constituting conditions of words. But on the other hand, the word is used as an adjective or adverb and we speak of being social, or of social action, or just of doing. Thus on the one hand society exists, and on the other hand society must be produced. And given this double nature to the way we see our topic there can be no fundamental origin of society or human behavior. There can be only the search for foundations. These foundations provide the antecedent conditions upon which all behavior is based. However, these foundations themselves are also based upon antecedent conditions. So the challenge to modern thought is to relate things that are the unconscious components of our existence to things that are the conscious components. It is the challenge to the papers in this conference.

Notes

1. Our terminology here does not parallel Thomas Kuhn's (1970) description of paradigm. This is because Kuhn is not talking about the human sciences, as we are. But there is another reason for our perspective. Concepts gain their meaning only in terms of other concepts with which they are in relationship—and so far, at least, this is what paradigm is—but in these relationships some concepts are embedded deeper in the structure than are others. These Gouldner (1970) would see as involved in the domain "assumptions" of a perspective and because of this position, they are less dependent for their meaning than those that depend upon them. But, then, this independence they have means they are not precisely definable.

2. It might be argued that Freud uses both terms in his theory and that the concepts of id and superego verify this. But for all the psychoanalytical

pointing in that direction, the theory never actually comes to biology. The id, for example, has no knowable somatic site; it is discussed in the theory in terms of its symbolic representations. Just exactly what blind hunger is or where it is located cannot be specified in Freud (see Ricoeur 1970). Quite simply, psychoanalysis loses its explanatory power when it ignores language or the symbol and focuses on the biology. Freud's dream of making a connection between mind and brain was not fulfilled; given the structure of his explanation it cannot be (see Ricoeur 1970, pp. 134–142).

3. This is, of course, an old argument, one that Marx fought with in his discussion of Feverbach. For Marx, man was a species being, his labor (because the theory is materialist) functions to give synthesis between man's natural being and his knowledge. That is, Marx conceives of thinking, as opposed to sensing, in terms of the nature of production: knowledge was produced. Habernas (1971, pp. 25–63) gives an excellent discussion of this.

4. But modern society does both. The same rational industrial society that promotes the health of the population can kill most of it in about 30 minutes. Rational techniques cut both ways.

5. Obviously this line has been blurring for some time: it is just that the recognition comes more easily as the society begins to think in terms of artificial replacement organs and their kin.

6. Theories of the midrange that Robert Merton (1945) proposes sociologists seek to construct, because they would be tested in empirical waters, never materialized. What the literature shows is much empirical research that is not theoretical in any rigorous sense of the term. Instead, modern research is justified in terms of a class of previous research.

7. Indeed, ecology, which seeks integration, suffers from this same crisis. Ecology is an approach to individual study, not a paradigm in Kuhn's sense of that term.

8. The tendency to supplement a theory of knowledge based on a sociology of, or a psychology of, or a biology of, knowledge with a philosophy of knowledge is most general in the human sciences. The attempt is to displace the threat of unconscious motive. Any sociology of knowledge must address the issue of unconscious social determinism, and this requirement on the approach means that questions of ideology and escape from ideological (and unconscious) determinants must be addressed; otherwise how can one claim transcendence and objectivity? In these circumstances, relative culture arguments can be damning; this leads to the tendency to seek support for objectivity in the womb of philosophy. It is this tendency that accounts for the current popularity of the work of Thomas Kuhn and the popularity of the concept of paradigm. Both Kuhn and the concept of paradigm offer relativity without threat, one theory is not wrong and another correct, or one ideological and another not. Both are simply paradigms. This makes things relative in much the manner the concept of subculture does.

9. Foucault offers an excellent discussion of the place of time in psychological models (primarily Freudian and developmental models) in *Mental Illness and Psychology* (1976, pp. 64–76).

10. In fact, from Marx forward, this constant winking in of the awareness of individual psychology has been a constituting feature of most sophisticated discussions of method. Weber, for instance, separates the individual's consciousness into that private and that shared, in terms of a separation he makes between value and fact.

11. Ethnomethodology depends, for its very existence, on the truth of this assumption (see Garfinkel 1967). Obviously so-called interpretive sociologies of phenomological sociologies do also.

12. *See* note 8.

13. This is no less true for the modern work of Piaget (1971) than it was for the work of Marx or Freud (who offered two integrated time scales: that of the development of species and that of the development of the individual). And Wilson's work again brings time back into the theoretical concern for behavior.

14. The Marxist eschatology integrates into its forms of being positive in terms of promise. Nothing about the theory can assure that the class revolution will be the last movement toward synthesis (see Foucault 1970, p. 320).

15. Indeed, many of the themes of social (especially sexual) repression in Freud no longer apply. The sexual revolution is social change and that change relates to personality and its illnesses. This is, of course, a problem common to approaches to mental illnesses that relate too closely to the processes of labeling in a society.

16. Sociobiology necessarily becomes a political concern because techniques exist to use the knowledge directly upon the population. In modern society social power now can be articulated on the population as a whole (in terms of energy or mass society) or upon the cell or even the gene. In this environment, it is inevitable that the gene pool should become a topic of political disclosure.

17. In his excellent structural history of genetics, Jacob argues that much of this theoretical integration is already proceeding (1973, pp. 299–324). But sadly, his hope for a total integration of all the levels from the physical to the social ignores the traditional epistemological issues surrounding the problem of language in consciousness.

References

Caplan, Arthur L. 1978. *The Sociobiology Debate.* New York: Harper and Row.

Durkheim, Emile. 1938. *The Rules of Sociological Method.* Chicago: The University of Chicago Press.

Foucault, Michel. 1970. *The Order of Things.* New York: Pantheon.

———. 1976. *Mental Illness and Psychology.* New York: Harper and Row.

Freud, Sigmund. 1949. *An Outline of Psycho-Analysis.* New York: N.W. Norton.

Garfinkel, Harold. 1967. *Studies in Ethnomethodology.* Englewood Cliffs, Prentice-Hall.

Gouldner, Alvin W. 1970. *The Coming Crisis of Western Sociology.* Garden City, N.Y.: Doubleday.

Habernas, Jorgen. 1970. *Knowledge and Human Interest.* Boston: Beacon Press.

Jacob, Francois. 1973. *The Logic of Life.* New York: Pantheon.

Kuhn, Thomas. 1970. *The Structure of Scientific Revolutions.* Chicago: University of Chicago Press.

Lovejoy, Arthur O. 1964. *The Great Chain of Being.* Cambridge, Mass.: Harvard University Press.

Merton, Robert K. 1945. *Social Theory and Social Structure.* New York: Free Press.

Piaget, Jean. 1971. *Biology and Knowledge.* Chicago: University of Chicago Press.

Ricoeur, Paul. 1970. *Freud and Philosophy.* New Haven and London: Yale University Press.

Rieff, Philip. 1961. *Freud: The Mind of the Moralist.* Garden City, N.Y.: Anchor Press/Doubleday.

Ryle, Gilbert. 1949. *The Concept of Mind.* New York: Barnes and Noble.

Wilson, Edward O. 1975. *Sociobiology: The New Synthesis.* Cambridge, Mass.: Harvard University Press.

———. 1978. *On Human Nature.* Cambridge, Mass.: Harvard University Press.

Worster, Donald. 1979. *Nature's Economy.* Garden City, N.Y: Anchor Press/Doubleday.

3

Darwinian Selection and Behavioral Evolution

Marcus Feldman and
L.L. Cavalli-Sforza

There was a time when a behavioral ecologist could present data on the behavior of insects, birds, rodents, or primates, and the science would be evaluated in terms of the experimental design, the novelty of the data collection technique, or the intrinsic fascination of the trait under study. It might even have been the relationship between the behavioral adaptation under study and some more or less cryptic chemical process that was the focus of attention. That time is past. A modern behavioral ecology seminar almost invariably begins with some theory couched in terms of what has become known as *sociobiology*. Operationally this entails the juxtaposition of some parameters of "fitness" and some measure of "relationship" among participants in the behavior to produce some conclusion about when the behavior should evolve. There usually follows some comparison of the observations with the theory, though rarely does this include an assessment of whether the assumptions of the theory are met.

To population geneticists, such as ourselves, the *notion* that theory should play a role in behavioral ecology is not distressing. After all, theory has been central to population genetics for more than seventy years. Even the idea that some interactions among members of a population are subject to genetic influences is not a major area of dispute. Mating behavior in *Drosophila*, for example, has been shown to be influenced by relative densities of participating genotypes [1] and there is a well-developed mathematical theory of assortative mating [2] and sexual selection [3] in which small numbers of genes are involved.

The aspects of sociology that have disturbed us are: (i). The theory has developed in a heuristic and verbal way (unparalleled in any other aspect of population biology) where careful specification of models, precise definition of parameters, and analysis of the robustness of conclusions to perturbation of assumptions have been sacrificed and replaced by a method of reasoning that we call the *inclusive fitness method*. This is a process of assigning fitness to an allele by accumulating contributions to its frequency

The research on which this chapter is based was supported in part by grants from the National Institutes of Health and the Department of Energy.

within generations and over generations. However, the conditions under which allele frequencies can replace genotype frequencies (a problem well studied in population genetics theory) are generally not met in the relevant models. We shall review some of our attempts to replace the inclusive fitness method with an analytical modeling approach. The latter, while technically more complex, places the theory of "altruism" among sibs, for example, in an orthodox population genetic context, as a particular case of Darwinian frequency dependent selection.

(ii). All published theory used to justify sociobiology is single-gene theory. There have been occasional forays into multigene applications, but these use the same heuristics as the one-locus approximations and are equivalent to single-gene treatments. Despite this, the ecology and evolution of behavioral traits whose biological basis is unknown and to a large extent untestable is explained in terms of single-gene theory. The extreme example of this is Dawkins' treatment [4], in which genes are freed of chromosomal constraints and seem to be given minds of their own, which they use to control the behavioral evolution of their carriers. Over the past ten years we have developed a theoretical approach to the evolution of traits whose transmission rules are complex [5]. A special focus of our interest has been the quantitative theory of cultural evolution, which can accommodate several of the phenomena modeled by sociobiologists. Our theory may be regarded as a more general approach that includes nongenetic alternatives to sociobiology's genetic theory of behavioral evolution.

Population Genetic Models of Kin Selection

The most intensively studied situations of kin selection involve *brood selection*. In broods (or more generally sibships) containing an individual of the "altruistic" genotype G_1 all members of that brood have higher Darwinian fitness than they would have had in the absence of the "altruist". By higher Darwinian fitness we mean either an increased chance of survival to adulthood, or increased fertility, or both. In the process of helping its sibs, the altruistic genotype G_1 suffers a fitness loss. The following remarks refer to brood selection defined in this way, although the framework can, in principle, be extended to more complex social interactions. For example, if G_1 acting as a parent increases the survival probability of its offspring over other parental genotypes, the situation is one of parent-to-offspring altruism, commonly termed parental care [6, 7]. There have been arguments as to whether this situation properly belongs in the sociobiology arena since it obviously involves direct differential selection on the parental genotypes. These arguments typify one of the major problems that arise when the inclusive fitness method is invoked, namely, what is direct and what is inclusive fitness. Careful model construction disposes of the argument by eliminating the need for the concept of inclusive fitness. We present here a summary of the diploid sib-to-sib case. This and the haplo-diploid cases (called sex-linked in population genetics) so central to the sociobiology of insects are discussed in detail elsewhere [6, 7, 8, 9]. We emphasize that our objective has been to understand the structure of a rigorously posed class of models rather than to apply the results. The models involve only the simplest assumptions and even when correctly analyzed must be viewed with extreme circumspection in relation to any actual behavior.

Consider a single gene with genotypes A_1A_1, A_1A_2, A_2A_2, which perform a certain behavior with probabilities h_{11}, h_{12}, h_{22}, respectively. The usual situation takes the behavior to be *altruism*, that is, the increase of fitness of one's sibs with a concomitant reduction of one's own fitness. The "penetrance" parameters h_{11}, h_{12}, h_{22} are usually taken to be 1, h, 0, respectively, with h a measure of the dominance of the trait ($1 \geq h \geq 0$). This more general parameterization was introduced by Uyenoyama and Feldman [8] for two alleles and generalized to the multiallele situation by Uyenoyama, Feldman, and Mueller [9].

We have suggested [6] two alternative fitness constructions, each of which incorporates the losses and gains due to "altruism." Suppose that in the present generation the frequencies of A_1A_1, A_1A_2, and A_2A_2 are u, v, and w, respectively. The fitness components for altruism can be summarized as in table 3–1.

In table 3–1, f_{ij} is the probability that A_iA_j has an altruistic sib. Generally it is assumed that $\beta > 0$ and $\gamma < 1$. The construction of f_{ij} makes use of elementary matrices of relationship. For the diploid sib-sib relationship table 3–2 is relevant.

In table 3–2, $p = u + v/2$ is the frequency of allele A_1 and the entries of the first row, for example are the probabilities that an individual of genotype A_1A_1 has a sib who is, respectively, A_1A_1, A_1A_2, A_2A_2. Such matrices are well known in population genetics (see Li [10]). Using the entries M_{ij} of the matrix M we write:

Table 3–1
Fitness Components for Genetic Altruism

	Fitness Loss *Due to Altruism Given*	*Fitness Gain* *Due to Altruism Received*
A_1A_1	γh_{11}	βf_{11}
A_1A_2	γh_{12}	βf_{12}
A_2A_2	γh_{22}	βf_{22}

Table 3-2
Diploid Sib-Sib Matrix M

$\text{sib}_1/\text{sib}_2$	A_1A_1	A_1A_2	A_2A_2
A_1A_1	$(u + v/4)^2/p^2$	$2v(u + v/4)/4p^2$	$v2/16p^2$
A_1A_2	$v(u + v/4)/4pq$	$[(v/4)^2 + (u + v/4)$ $(w + v/4)]/pq$	$v(w + v/4)/4pq$
A_2A_2	$v^2/16q^2$	$2v(w + v/4)/4q^2$	$(w + v/4)^2/q^2$

$$f_{11} = h_{11}M_{11} + h_{12}M_{12} + h_{22}M_{13}$$
$$f_{12} = h_{11}M_{21} + h_{12}M_{22} + h_{22}M_{23}$$
$$f_{22} = h_{11}M_{31} + h_{12}M_{32} + h_{22}M_{33}$$

This completes the construction of the fitness components. It remains to put them together. In [6] we suggested an additive and a multiplicative fitness composition:

Additive: A_1A_1: $\phi_{11} = 1 - \gamma h_{11} + \beta f_{11}$

             A_1A_2: $\phi_{12} = 1 - \gamma h_{12} + \beta f_{12}$

             A_2A_2: $\phi_{22} = 1 - \gamma h_{22} + \beta f_{22}$

Multiplicative: A_1A_1: $\phi_{11} = (1 - \gamma h_{11})(1 + \beta f_{11})$

             A_1A_2: $\phi_{12} = (1 - \gamma h_{12})(1 + \beta f_{12})$

             A_2A_2: $\phi_{22} = (1 - \gamma h_{22})(1 + \beta f_{22})$ (3.1)

Of course it is assumed here that A_1 and A_2 have no effects on fitness other than the losses and gains described.

At birth the frequencies of A_1A_1, A_1A_2, A_2A_2 are p^2, $2pq$, q^2, respectively. After selection we have the frequencies in the next generation [6, 8].

$$Tu^1 = p^2\phi_{11}$$
$$Tv^1 = 2pq\phi_{12}$$
$$Tw^1 = q^2\phi_{22}$$ (3.2)

where $T = p^2\phi_{11} + 2pq\phi_{12} + q^2\phi_{22} = 1 + (\beta - \gamma)(p^2h_{11} + 2pq\,h_{12} + q^2h_{22})$. It should be stressed that the additive and multiplicative constructions are two among a very large number of reasonable models that might be used.

The usually considered parameter set is $h_{11} = 1$, $h_{12} = h$, $h_{22} = 0$. Thus, A_2A_2 does not perform altruistically, A_1A_2 does so with probability h, and A_1A_1 does so with probability one. Thus, A_1 has been called the *altruistic gene*. The first treatment of models of this type was made by Hamilton [11], who focused on the haplo-diploid (or sex-linked) case and its relevance to the social *Hymenoptera*. Hamilton used an approach that turns out to be equivalent to the *additive* structure and explicitly made the Hardy-Weinberg approximation $u = p^2$, $v = 2pq$, $w = q^2$. This approximation allows equation 3.2 to be analyzed as if it were a one-dimensional relationship between the gene frequency p (now) and p' in the next generation. Thus if $p' > p$ it is said that the altruistic gene increases. This one-dimensional approximation has the further advantage (in the eyes of its advocates) that it allows more complicated patterns of relationship to be studied, patterns whose exact analytic formulation is out of the question at present. It is this one-dimensional approximation and the conclusions drawn from it, assumed to apply to the altruistic gene A_1, that we call the *inclusive fitness method* to distinguish it from the analysis and results obtained from the exact analysis of the frequency dependent recussion 3.2, which is two-dimensional.

The inclusive fitness method leads to Hamilton's rule for the increase of the "altruistic gene." This is usually stated in the form that the altruistic gene increases if

$$\beta r_H > \gamma \qquad (3.3)$$

where β and γ are as in (3.2) and r_H is some measure of the relationship between donor and recipient of the altruism. For example, in the diploid case specified by equations 3.2 the condition for initial increase would be $\beta/2 > \gamma$, according to the standard treatments. In this case r_H is the probability that the gene chosen randomly from one sib is identical by descent to one chosen from the other sib. In the haplo-diploid case sister-to-sister altruism produces $r_H = 3/4$ by this definition. Thus the gain-to-loss ratio sufficient for Hamilton's rule to apply in this case is less than that in the diploid case. This has been used to explain the repeated evolution eusociality in the social Hymenoptera. It is essential in our assessment of Hamilton's rule to understand the assumptions that have been made in its derivation. Many of those who followed Hamilton have not only been less careful than he was in determining where approximations are made, but also more willing to extrapolate from the idealized model situation to actual behaviors in higher animals. Since the assumptions made are tacit and usually ignored it seems worthwhile to spell them out.

Assumption 1. One allele is defined as the altruistic gene. For example, if this is A_1 then the assumption for equations 3.2 to determine the increase of A_1 when it is rare is $h_{12} > h_{22}$. If this condition is reversed, that is, $h_{12} < h_{22}$, then the inequality (3.3) must be reversed also. This is a significant consideration in light of the fact that there are no known examples for which the ordering of h_{11}, h_{12}, h_{22} can even be guessed at.

Assumption 2. If $h_{12} > h_{22}$ then formula 3.3 is the condition for fixation of A_1. This is usually a tacit assumption, although it has been made explicitly in some studies. These are several ways in which this assumption can break down. If $h_{11} > h_{12} > h_{22}$ so that formula 3.3 determines the local stability of fixation in A_1, it does not logically follow that A_1 will fix. We showed [6] that interior polymorphic equilibria may intervene to prevent initially rare A_1 from proceeding to fixation even when formula 3.3 determines its initial increase. The detailed analyses of these polymorphisms are presented in [8] and [9]. These studies also pursue the analyses when $h_{11} < h_{12} > h_{22}$, in which case another class of equilibria emerges.

Assumption 3. Inequality 3.3 applies for cross-sex "altruism." In the sister-to-brother haplo-diploid case we showed [6] that the rule fails. This led us to seek some more general interpretation of formula 3.3 and in [8] and [9] we show that initial increase is described by (3.3) in all cases if r_H is the regression of the recipient's additive genotype value on that of the donor. These studies also produce general conditions for the polymorphic equilibria in multiallele models.

Assumption 4. The fitness construction is additive. In [6] we showed that under the multiplicative fitness arrangement Hamilton's rule formula 3.3 is not the condition for initial increase.

The inclusive fitness method, by reducing the models to gene frequency recursions, hides most of these assumptions, and in particular gives a misleading picture of the dynamics of the genotype frequencies in the population. Complete multidimensional analyses reveal the structural complexities of Hamilton's original models, which the more recent development of sociobiology, in its rush to obtain simple consequences of genetic determination, has managed to camouflage.

Quantitative Theory of Cultural Evolution

No known social behavior satisfies the simple model of determination and transmission described in the previous section. The role of learning from parents and other members of a population group has become a focus of research in animal behavior (see for example the recent review by Mundinger [12]). In human behavioral research the approach has usually been a static one in that a statistical partitioning of variation into genetic and other causes is made in the absence of a dynamic evolutionary context. Our work in recent years has aimed to develop such an evolutionary framework in which cultural transmission plays the central role.

The following example illustrates how a theory based on nongenetic transmission can allow the evolution of individually disadvantageous traits. Consider a trait that takes two possible values in a population, say $+$ and $-$. The transmission considered is vertical (although other types of transmission are considered elsewhere, see [5]) and can be represented by table 3-3, analogous to a segregation table in genetics. In table 3-3, b_3, for example, is the probability that in offspring of the $+ \times +$ mating is $+$. A special case analogous to biological transmission (in haploids) would take $b_3 = 1, b_2 = b_1 = 1/2, b_0 = 0$. In this case any selective disadvantage to $+$, for example in survival probability, would result in eventual loss of $+$.

Now suppose that in table 3-3 one $+$ parent is sufficient for the transmission of $+$ to the offspring, that is, $b_3 = b_2 = b_1 = 1; b_0 = 0$. Then $+$ will be preserved in the population as long as the chance of $+$ offspring to survive is greater than 50 percent of that of $-$ offspring.

The example suggests that by expanding our view of the rules of transmission the paradox of the evolution of individually disadvantageous traits, that led to the predominance of kin selection theory, can be overcome.

In examples like that of table 3-3 transmission and viability act in opposite directions. We can expand the meaning of these two concepts into *cultural selection* on the one hand and *natural selection* on the other. The latter is clearly defined in terms of the different chances for the various types in a population to be represented in the next generation. The former, however, is a more nebulous concept. It might be summarized in certain cases by the b_i. In the human context, *cultural selection* describes an amalgamation of the process of *transmission* of the trait, *decision making* concerning acceptance of the trait, and the process of *acceptance* of the trait by a recipient.

Three examples illustrate the complexity of the cultural selection pro-

Table 3-3
Transmission of a Two-Valued Trait

Mother	Father	Offspring	
		$+$	$-$
$+$	$+$	b_3	$1 - b_3$
$+$	$-$	b_2	$1 - b_2$
$-$	$+$	b_1	$1 - b_1$
$-$	$-$	b_0	$1 - b_0$

cess and its interaction with natural selection. In the case of drug abuse, ignorance of the danger, peer pressure or "promotion," and genuine addiction may all mediate against correct perception of the danger associated with adoption of the trait, which may result in loss of Darwinian fitness. Female circumcision is a practice fraught with danger to the health of the subject. Yet even today the custom is common in many areas of the world under Arabic influence. The trait is maintained as a cultural practice possibly because of a belief prevalent in such cultures as Islam that circumcision guarantees the morality of women. A third example is the use of birth control. This practice clearly reduces the fitness of its practitioners in the Darwinian sense, but at least Western society has been selected for in the *cultural* sense over the past one hundred years.

What must be stressed in these examples is that no biological transmission mechanism has to be invoked. Nor does one have to invent just-so stories to explain the "adaptive value" of these customs (see also Gould and Lewontin [13]). Once the complexity of the transmission is admitted we are forced to orient more toward the processes of cultural selection rather than of natural selection.

Conclusion

Hamilton's original treatment of what is now called kin selection was, of course, much more limited in scope than what is now called sociobiology. His conclusions from his approximations were largely correct, but the departures from his results, found using exact methods, reveal a surprisingly rich theoretical structure. They also reveal that subsequent verbal and approximate treatments must be scrutinized most carefully so that hidden assumptions can be explicated. As we have seen, some of these treatments can be made more rigorous. But in so doing they lose the quality of simplicity and unidimensionality that has turned out to be so appealing to sociobiology.

Alternative evolutionary models, of which the one described here is but the simplest example, introduce tremendous flexibility through transmissions assumptions. These can resolve the paradox of the maintenance of individually disadvantageous traits by shifting the emphasis away from Darwinian fitness consideration to the transmission process we call cultural selection.

References

1. Ehrman, L. and P.A. Parsons. 1976. *The Genetics of Behavior.* Sunderland, Mass.: Sinauer Associates.

2. Crow, J.F. and M. Kimura. 1970. *An Introduction to Population Genetics Theory.* New York: Harper and Row.
3. O'Donald, P. 1980. *Genetic Models of Sexual Selection.* New York: Cambridge University Press.
4. Dawkins, R. 1976. *The Selfish Gene.* New York: Oxford University Press.
5. Cavalli-Sforza, L.L. and M.W. Feldman. 1981. *Cultural Transmission and Evolution—A Quantitative Approach.* Princeton, N.J.: Princeton University Press.
6. Cavalli-Sforza, L.L. and M.W. Feldman. 1978. Darwinian selection and "altruism." *Theor. Pop. Biol.* 14:268–280.
7. Feldman, M.W. and L.L. Cavalli-Sforza. 1981. Further remarks on Darwinian selection and "altruism." *Theor. Pop. Biol.* 19:251–260.
8. Uyenoyama, M.K. and M.W. Feldman. 1981. On relatedness and adaptive topography in kin selection. *Theor. Pop. Biol.* 19:88–123.
9. Uyenoyama, M.K., M.W. Feldman, and L.D. Mueller. Population genetic theory of kin selection: I. Multiple alleles at one locus. *Proc. Nat. Acad. Sci.* In press.
10. Li, C.C. 1955. *Population Genetics.* Chicago: University of Chicago Press.
11. Hamilton, W.D. 1964. The genetical evolution of social behavior. I, II. *J. Theor. Biol.* 7:1–52.
12. Mundinger, P.C. 1980. Animal cultures and a general theory of cultural evolution. *Ethology and Sociobiology* 1:183–223.
13. Gould, S.J. and R.C. Lewontin. 1979. The spandrels of San Marco and the Panglossian paradigm: A critique of the adaptationist programme. *Proc. Roy. Soc. Lond. B.* 205:581–598.

4

Endorphins: Endogenous Control of the Perception of Pain

G.C. Davis, M.S. Buchsbaum, and W.E. Bunney, Jr.

The purpose of this chapter is to review the evidence that endogenously produced opiate-like substances (endorphins) play a role in the regulation of pain perception. Endorphins are the generic name for a series of small peptides that have been found in the pituitary, brain, and other tissues. The generic name, *endorphins,* is a contraction of "endogenous morphines," so named because these peptides have actions similar to narcotic analgesics.

The discovery of endorphins was made possible by the successful search for molecular mechanisms that account for the actions of drugs. Ehrlich (1913) hypothesized that there were specific chemical groupings in cells with which certain drugs reacted specifically. Ehrlich called these chemical groupings *receptors.* Langley (1909) investigated the action of curare and named the site of action in muscle *receptive substance.* Today the term *receptor* is used to describe a macromolecule with which a drug interacts to produce its characteristic biological effect. For a number of years investigators have postulated the existence of specific receptors on nerve cells that mediate the actions of narcotic analgesics. The recent development of techniques for isolating binding of drugs to cell fractions and the demonstration of receptor characteristics for that binding were necessary for the discovery of the opiate receptor and subsequently of the endorphins.

Pert and Snyder (1973), Terenius (1973), and Simon, Hiller, and Edelman (1973) isolated in vitro the opiate receptor, a macromolecular constituent of neurons that possessed the property of specific binding to opiates. As no one believed that evolution had endowed man with an opiate receptor in order that drugs could be administered, for example, to suppress pain, it was assumed that the body must synthesize substances whose normal function was to act at these receptor sites. A search for such endogenous substances (called natural ligands because of their binding to receptors) culminated in the discovery by Hughes et al. (1975) of leucine- and methionine-enkephaline, two 5-amino-acid peptides with opiate-like activity. Subsequently other substances have been demonstrated in cerebrospinal fluid (Terenius and Wahlström 1975), brain (Frederickson et al. 1976), and peripheral blood (Pert, Pert, and Tallman 1976). The amino-acid sequence of the endorphin methionine-enkephalin was found within the hormone beta-lipotropin (a 91-amino-acid protein). Enzymatic digests of beta-lipo-tropin (B-LPH) produced other substances that possess opiate-like activity, the best known of which is beta-endorphin (amino-acid sequence 61–91 of BLP).

The small opiod peptides methionine- and leucine-enkephalin are thought to be neurotransmitters and have been localized in specific brain regions. Beta-endorphin, on the other hand, has not only been associated with a neuronal tract in the brain but is also located in the anterior pituitary, where it may be released into the blood to act as a hormone. Robert and Herbert (1977) and Mains, Eipper, and Ling (1977) demonstrated that there is a large prohormone (some call pro-opiocortin) containing the amino-acid sequences for both beta-lipotropin and adrenocorticotropic hormone (ACTH). Krieger, Liotta, and Li (1977) demonstrated that B-LPH and ACTH are released from the pituitary in equimolar amounts in diseases of pituitary hyperfunction such as Cushing's disease. Thus beta-endorphin and possibly other opiod peptides may function as hormones, their target effects occurring at a location distant from that of their production.

Discovery of endorphins stimulated a search for the physiological role of these opiate-like substances. If the pharmacological effects of narcotic analgesics provide the best model for the physiological functions of endorphins, then since morphine alters mood, pain appreciation, sleep, respiration, and the release of pituitary hormones, these functions are prime candidates for endorphin activity. In this chapter we will investigate whether endorphins play a role in the perception of pain in man.

Opiate Receptors, Endorphins, and the Neuroanatomy of Pain

Where are opiate receptors and endorphins located in the central nervous system? As one would expect for substances suspected of influencing pain appreciation, the neuroanatomical distribution of opiate receptors and endorphins is found in central nervous system (CNS) areas key to nociceptive information processing; spinal dorsal horn and supraspinal sites in the periaqueductal gray matter of the midbrain, the nucleus gigantocellularis of the pontine reticular formation, and the medial thalamus. These neuroanatomical locations are ideally suited to inhibit nociceptive impulses transmitted through the spinothalamic tract. For details of the neuroanatomy and neurophysiology of opiate receptors see Pert et al. (in press).

Are Endorphins Analgesic?

Beta-endorphin has been administered to man in three studies in order to test its analgesic potency. Intravenous injection produced analgesia in two of three cancer patients experiencing pain (Catlin et al. 1977) while intraventricular (IVT) administration produced analgesia in three patients with intractable pain (Hosobuchi and Li 1978 *a, b*). Furthermore, the intravenous administration of naloxone reversed the analgesia produced in the later study. In another study of intractable pain in cancer patients, Foley et al. (1978) found that intravenous doses of 5 and 10 mg failed to bring about relief while IVT doses of 7.5 mg produced analgesia and marked drowsiness. The parallel CNS distribution of enkephalins and opiate receptors as well as the analgesia produced in animals by direct application of endorphins to brain regions (by microintophoresis), support the hypothesis that endorphins function physiologically to reduce pain.

Endorphins: A Tonic or Phasic Role?

If endorphins play a tonic role in inhibiting pain perception, then narcotic antagonists should increase pain by blocking or reversing endorphin-produced insensitivity. Jacob, Tremley, and Colombel (1974), Frederickson et al. (1976), and Bernston and Walker (1977) have shown that naloxone produces hyperalgesia in mice and rats.

Lasagna (1965) reported data that suggest that naloxone in certain doses increases pain in humans. While a number of human studies have been unable to demonstrate naloxone-induced hyperalgesia (Grevert and Goldstein 1977, 1978; El-Sobky, Dostrovsky, and Wall 1976), Buchsbaum, Davis, and Bunney (1977) not only reported hyperalgesia following naloxone but presented data suggesting that individual differences in pain sensitivity are mediated by endorphins: naloxone-induced hyperalgesia was found predominantly in pain-tolerant subjects. This finding has been replicated in guinea pigs (Satoh et al. 1979). Levine et al. (1978) have reported that naloxone increased pain following dental extractions.

Why has naloxone induced hyperalgesia in some studies and not in others? Demonstration of pharmacological analgesia using acute experimentally induced pain is difficult, and since the reported hyperalgesic effects of naloxone appear to be small, sophisticated measurement techniques and stimulus technology are necessary. Nevertheless, it is doubtful that these methodological issues alone explain the failure of investigators to demonstrate such hyperalgesia. An additional explanation is that endorphin-mediated pain insensitivity may be phasic rather than tonic. Frederickson, Burgis, and Edwards (1977) found that naloxone damped the diurnal variation in pain sensitivity in mice. Similarly in man, Davis, Buchsbaum, and Bunney (1978) demonstrated diurnal variation in pain sensitivity in which human subjects were found more insensitive to pain in the morning than in the afternoon. Furthermore, naloxone-induced hyperalgesia was found in subjects tested in the morning and not in the afternoon, supporting the hypothesis that endorphin release is phasic. In animals, endorphins are released by some stresses and not others, and by different degrees of stress, suggesting that the amount or type of pain or stress may be relevant to whether there is a phasic release.

Stimulation-Produced Analgesia

Focal electrical stimulation of the brain produces analgesia in animals and in man (Reynold 1969; Mayer et al. 1971; Mayer and Liebeskind 1974; Oliveras et al. 1974) that is reversible by naloxone (Oliveras et al. 1977). Tolerance develops to the analgesic effects of brain stimulation, and cross-tolerance has been demonstrated between morphine and brain stimulation (Mayer and Hayes 1975).

Electrical stimulation of the periventricular gray matter at the level of the posterior commissure in human patients produces analgesia that is reversed by naloxone (Adams 1976; Hosobuchi, Adams, and Linchutz 1977). Furthermore, focal stimulation of a medial thalamic site adjacent to the third ventricle and periaqueductal gray matter produced increases in beta-endorphin-like immunoreactivity in human cerebrospinal fluid (CSF) (Akil et al. 1978) and ventricular fluid (Hosobuchi et al. 1979). Analgesia produced by focal electrical stimulation in periaqueductal gray matter is partially reversed by the narcotic antagonist naloxone (Akil, Mayer, and Liebeskind 1976). These data suggest that endorphins may be released by the electrical stimulation of specific brain regions, and that they mediate the analgesia produced by such stimulation.

Endorphins in Cerebrospinal Fluid

Focal stimulation of sites of the human brain raises levels of beta-endorphin-like immunoreactivity in the CSF (Akil et al. 1978). In a study of chronic pain syndromes, von Knorring et al. (1978) reported that patients with high levels of opiate-binding material in their CSF had greater pain tolerance than those with low levels. Thus these authors suggest that endorphins are physiological regulators of pain threshold and tolerance. This group has also reported elevations of endorphins in human CSF during psychotic states (Lindström et al. 1978). Furthermore, this group has reported a correlation between depth of depression and elevations in opiate binding in the CSF in chronic pain patients (Almay et al. 1978), suggesting a link connecting endorphins, chronic pain, and depression that we discuss in greater detail below.

Acupuncture Analgesia

Pomeranz and Chiu (1976) report that naloxone blocks electroacupuncture analgesia in mice and thus implicate endorphin release as a mechanism responsible for this type of analgesia. Mayer, Price, and Rafri (1977) have reported that acupuncture analgesia appears to involve activation of endorphin release in man. In subjects selected for successful acupuncture analgesia, naloxone reduced pain thresholds to the level of the placebo control group. These authors suggest that it is unlikely that an endogenous agent is released at the acupuncture site since vascular occlusion did not interfere with this analgesia. Rather they implicate a CNS site of action, particularly since acupuncture was ineffective when the acupuncture point or peripheral nerve was anesthetized with procaine. Sjölund, Terenius, and Eriksson (1977) report that several of nine patients with chronic pain undergoing electroacupuncture had elevated levels of endorphins in the CSF.

Hypnotic Analgesia

Goldstein and Hilgard (1975) were intrigued by the similarities between hypnotic analgesia and opiate analgesia and tested the hypothesis that hypnotic analgesia might be mediated by endorphins. However, they found no effect of naloxone on hypnotically produced analgesia to ischemic pain in three subjects. Barber and Mayer (1977) studied normal subjects and subjects undergoing dental procedures with hypnotic analgesia. These subjects also failed to demonstrate reversal of the hypnotic analgesia by naloxone.

Placebo Analgesia

Levine, Gordon, and Fields (1978, 1979) suggest that analgesia produced by placebo might be a consequence of release of endorphins. Patients undergoing mandibular third molar extractions were divided into placebo responders and nonresponders based on their pain report after placebo. Naloxone given subsequently caused hyperalgesia in the placebo responder group and not in the nonresponders. Mihic and Binkert (1978) were unable to demonstrate reversal of placebo analgesia with 0.4 mg I.V. naloxone.

Congenital Insensitivity to Pain

Dehen et al. (1978) studied a patient with congenital insensitivity to pain. Naloxone caused a large fall of the nociceptive reflex threshold in this patient not found in eight controls. Injection of saline failed to effect reflex threshold. These authors suggest that this patient's insensitivity might be a consequence of endorphin excess. Further studies of possible genetic or congenital impairment of endorphin systems might provide valuable information on endorphin functions.

Pain, Psychiatric Illness, and Endorphins

Perceptual dysfunction has been a theme of investigations of the psychology of schizophrenia (for example, Venables 1973, Garmezy 1978). Abnormal visual, auditory, and olfactory sensation in schizophrenia involves not only sensory thresholds but also the integration of sensory information as well. It is therefore not surprising that the normal perception of pain appears disturbed in schizophrenia. Clinical descriptions abound of abnormalities of pain perception in schizophrenia.

Even severe and acute clinical conditions usually experienced as exquisitely painful may be less strongly appreciated in psychotic patients. Marchand et al. (1969) reported that pain did not accompany acute perforated peptic ulcer, acute appendicitis, or fracture of the femur in 21, 37, and 42 percent of seventy-nine psychotic patients presenting with these surgical disorders. In an earlier study Marchand (1955) found that 82.5 percent of psychotic patients with myocardial infarction presented without pain.

Pain insensitivity has been reported not only in clinical conditions but also in experimental studies of pain in schizophrenic patients. Malmo and Shagass (1949) found that a subgroup of schizophrenics are less able to discriminate among intensities of thermal stimulation than are normals or other psychiatric diagnostic groups. Malmo, Shagass, and Smith (1951) found that chronic schizophrenics have higher thresholds for thermally induced discomfort. Hall and Stride (1954) found diminished reactivity to painful stimuli. Sappington (1973) also described pain insensitivity among "process" in comparison to "reactive" schizophrenics.

We studied seventeen off-medication hospitalized schizophrenics and seventeen age- and sex-matched normal controls (Davis et al. in press). Schizophrenics were less pain sensitive than were the controls. This significant insensitivity is also reflected in a pattern of somatosensory-evoked potentials that we find correlated with insensitivity in normals (Buchsbaum 1975; Buchsbaum, Davis, and Bunney 1977). It is tempting to speculate that the pain insensitivity in a subgroup of schizophrenics might be secondary to an elevation of endorphins. Indeed we might further speculate that the beneficial clinical effects of naloxone in schizophrenics (reported by Gunne, Lindström, and Terenius 1977; Emrich et al. 1977; Watson et al. 1978) could be based in part on naloxone's action on abnormal sensory mechanisms in schizophrenia.

We have also studied the chronic effect of naltrexone, a long-acting, oral, narcotic antagonist, in five patients with schizophrenia. While the number of patients is too small to assess the clinical effects on psychotic symptomatology, the increase in pain sensitivity was striking (Davis et al. in press). This increase in pain sensitivity after naltrexone was also demonstrated by a sensitizing effect in somatosensory-evoked potentials in these schizophrenics. Cortical potentials evoked by electrical shocks increased in amplitude following naltrexone. This increase in amplitude is associated with hyperalgesia in the peak at 120 milliseconds (Buchsbaum, Davis, and Bunney 1977). Somatosensory-evoked potentials, abnormally small in schizophrenics (Buchsbaum 1977) were doubled or even tripled in size when recorded while patients were being treated with naltrexone. This evoked potential amplitude increase with naltrexone is compatible with our earlier report of a sensitizing effect of naloxone (Buchsbaum, Davis, and Bunney 1977).

Pain and Affective Illness

The physiological, behavioral, and affective effects of chronic pain are quite similar to the symptoms of depression. Patients with chronic pain, as do depressed patients, typically report loss of appetite, sleep disturbances, decreased libido, inability to concentrate or to take an interest in things, and loss of ability to function at work.

Patients with affective illness share with schizophrenic patients an insensitivity to experimental stimuli. Hemphill, Hall, and Crookes (1952), and Hall and Stride (1954) using heat pain; Merskey (1965b) using a pressure algometer; and von Knorring (1975) using electrical stimulation found that depressed patients have increased pain thresholds and pain tolerance when contrasted to normals. We have found that affectively ill patients are significantly more "analgesic" to experimental stimuli than are controls (Davis, Buchsbaum, and Bunney 1979). Male depressed patients were significantly more analgesic than female depressed patients or controls. We have also been able to distinguish between bipolar and unipolar affectively ill patients; bipolars were analgesic on a pain measure associated with pharmacological analgesia while unipolars were analgesic on a pain measure that correlates better with suggestion or other cognitive manipulations. This suggests that the analgesia seen in these illnesses may be mediated by different mechanisms. Pain appreciation in affective illness presents an interesting contrast to schizophrenia. While both groups are less sensitive to experimental pain, affectively ill patients report more somatic distress (for example, headaches, backaches). Indeed, such symptoms are often the presenting complaint of depression (see Merskey 1965a; Devine and Merskey 1965).

Interestingly, as with schizophrenia, there is indirect and preliminary evidence suggesting that endorphins may play a role in both the depressive mood and pain insensitivity of affective illness. Cyclazocine, a narcotic antagonist, has been reported to have antidepressant activity (Fink et al. 1970). While we have failed to document mood improvement in affectively ill patients administered the short-acting parenteral antagonist naloxone (Davis et al. 1977), endorphins may be involved in the analgesia found in some affectively ill patients (Davis, Buchsbaum, and Bunney in 1979). Terenius et al. (1976) found an elevated CSF endorphins in manic-depressive patients. Furthermore, Almay et al. (1978) reported that patients with psychogenic pain syndromes have elevated opiate receptor-binding material in their cerebrospinal fluid.

It is intriguing that naloxone increased separation distress and vocalization in guinea pig infants, while low doses or morphine suppressed these behaviors (Herman and Panksepp 1978), since maternal separation and the distress associated with it has been used as an animal model of depression.

Future Directions and Strategies

Though the discovery of endorphins occurred only several years ago, an enormous amount of basic and clinical data has accumulated since that time that has yet to be integrated into research designs. For example, naloxone's potency as an antagonist varies in various physiological preparations (Lord et al. 1977), and in fact three receptors have been postulated from narcotic studies in the chronic spinal dog (Martin et al. 1976; Gilbert and Martin 1976). Most clinical research has depended on the use of narcotic antagonists that appear to have potent effects at only one of the putative opiate receptors. These receptors may have different affinities for various opiate analgesics and endorphins, potentially explaining some of the discrepancies in the studies done to date. The discovery of the specific endorphin that acts as each receptor and the development of specific antagonists for each receptor type will certainly be of help in characterizing the physiological functions of these different receptors.

As behavioral effects of narcotic antagonists and endorphins become specified, traditional pharmacological tools such as dose-response studies should be utilized.

Metabolic strategies to uncover endorphin action pursued at present include the synthesis of compounds that irreversably bind to the opiate receptor (for example, Craviso and Mucasshio 1976) and the use of specific peptidases to slow endorphin degradation.

Certainly all endorphins have not been discovered. In addition to identifying additional peptides, there may be nonpeptide opiate-binding compounds in man. Furthermore, Klee, Zioudron, and Streaty (1979) found that enzymatic digests of wheat glutin bind to opiate receptors (thus the term *exorphins* enters our lexicon).

Knowledge of endorphin mechanisms of pain suppression may help answer several questions that have bedeviled pain researchers to date. What are the differences in mechanisms regulating (or failing to regulate) acute and chronic pain? What are the body's homeostatic responses to chronic sensory stimulation and mechanisms by which the CNS perpetuates the pain in the absence of peripheral stimulation? Finally, can endorphin-dependent neuronal mechanisms explain wide individual differences in susceptibility or response to pain?

Summary

In this chapter, we have reviewed human studies that suggest a possible physiological role for endorphins in pain suppression. Endorphins appear to function as neurotransmitters and as endocrine factors. Endorphins and opiate receptors are located in the central nervous system in areas key to pain information processing. When endorphins were administered to cancer patients some relief of intractable pain was demonstrated. It appears that endorphins may play a tonic role in pain suppression, but it is a weak action. Phasic release of endorphins appears to play a larger role.

Electrical stimulation of certain regions of the brain produces an elevation in endorphin levels in the cerebrospinal fluid. This analgesia is partially reversible by naloxone, additional proof that endorphins function to inhibit pain. Preliminary evidence suggests that acupuncture but not hypnosis is mediated by endorphins; furthermore, analgesia produced by placebos may in part be mediated by endorphins. Alterations in pain appreciation in affective illness and schizophrenia may involve endorphins.

Endorphinergic neurons and possibly endorphinergic humoral mechanisms appear to play a role in a variety of pharmacological and nonpharmacological analgesic interventions.

We suggest that endorphin research holds promise for understanding pain mechanisms not because it is the only, or even the most important, neuronal system mediating the organisms' capacity for pain suppression, but because the identification of opiate receptor and the discovery of opiate peptides have brought with them a new technology and—more importantly—new excitement to this field.

References

Adams, J.E. Naloxone reversal of analgesia produced by brain stimulation in the human. *Pain* 2:161–166, 1976.

Akil, H.; Mayer, D.J.; and Liebeskind, J.S. Antagonism of stimulation-produced analgesia by naloxone, a narcotic antagonist. *Science* 191: 961–962, 1976.

Akil, H.; Richardson, D.E.; Barchas, J.D.; and Li, C.H. Appearance of β -endorphin-like immunoreactivity in human ventricular cerebro-spinal fluid upon analgesic electrical stimulation. *Proc. Natl. Acad. Sci.* (USA) 75:5170–5172, 1978.

Almay, B.G.L.; Johansson, F.; von Knorring, L.; Terenius, L.; and Wahlström, A. Endorphins in chronic pain. I. Differences in CSF endorphin levels between organic and psychogenic pain syndromes. *Pain* 5:153–162, 1978.

Barber, J., and Mayer, D. Evaluation of the efficacy and neural mechanisms of a hypnotic analgesic procedure in experimental and clinical dental pain. *Pain* 4:41–48, 1977.

Buchsbaum, M.S. Averaged evoked response augmenting/reducing in schizophrenia and affective disorders. In Freedman, D.X. (ed.), *Biology of the Major Psychoses.* New York: Raven Press, pp. 129–142, 1975.

———. The middle evoked response components and schizophrenia. *Schizophr. Bull.* 3:93–104, 1977.

Buchsbaum, M.S.; Davis, G.C.; and Bunney, W.E., Jr. Naloxone alters pain perception and somatosensory evoked potentials in normal subjects. *Nature* 270:620–622, 1977.

Berntson, G.C., and Walker, J.M. Effect of opiate receptor blockade on pain sensitivity in the rat. *Brain Res. Bull.* 2:157–159, 1977.

Catlin, D.H.; Hui, K.K.; Loh, H.H.; and Li, C.H. Pharmacologic activity of β -endorphin in man. *Commun. Psychopharmacol.* 1:493–500, 1977.

Craviso, G.L., and Musacchio, J.M. Opiate receptor: Irreversible inactivation by an alkylating local anesthetic. *Life Sciences* 18:821–827, 1976.

Davis, G.C.; Buchsbaum, M.S.; and Bunney, W.E., Jr. Naloxone decreases diurnal variation in pain sensitivity and somatosensory evoked potentials. *Life Sci.* 23:1449–1460, 1978.

———. Analgesia to painful stimuli in affective illness. *Amer. J. Psychiat.* 139 (September):1148–1151, 1979.

Davis, G.C.; Buchsbaum, M.S.; van Kammen, D.L.; and Bunney, W.E., Jr. Analgesia to pain stimuli in schizophrenics reversed by naltrexone. *Psychiatry Res.* in press.

Davis, G.C.; Bunney, W.E., Jr.; DeFraites, E.G.; Kleinman, J.E.; van Kammen, D.P.; Post, R.M.; and Wyatt, R.J. Intravenous naloxone administration in schizophrenia and affective illness. *Science* 197:74–77, 1977.

Dehen, H.; Willer, J.C.; Prier, S.; Boureau, F.; and Cambier, J. Congenital insensitivity to pain and the morphin-like analgesic system. *Pain* 5:351–358, 1978.

Devine, R., and Merskey, H. The description of pain in psychiatric and general medical patients. *J. Psychosom. Res.* 9:311–316, 1965.

Ehrlich, P. Chemotherapeutics: Scientific principles, methods and results. *Lancet* 2:445, 1913.

El-Sobky, A.; Dostrovsky, J.D.; and Wall, P.D. Lack of effect of naloxone in pain perception in humans. *Nature* 263:783–784, 1976.

Emrich, H.M.; Cording, C.; Pirée, S.; Kolling, A.; v. Zerssen, D.; and Herz, A. Indication of an antipsychotic action of the opiate antagonist naloxone. *Pharmakopsychiatry* 10:265–270, 1977.

Fink, M.; Simeon, J.; Itil, T.M.; and Freedman, A.M. Clinical antidepressant activity of cyclazocine—a narcotic antagonist. *Clin. Pharmacol. Ther.* 11:41–48, 1970.

Foley, K.M.; Kaiko, R.F.; Inturrisi, C.E.; Posner, J.B.; Li, C.H.; and Houde, R.W. Intravenous and intraventricular administration of beta-endorphin in man: Preliminary studies. *Pain Abstracts,* vol. 1, Second World Congress on Pain, Montreal, Canada, 1978.

Frederickson, R.C.A.; Burgis, V.; and Edwards, J.D. Hyperalgesia induced by naloxone follows diurnal rhythm in responsivity to painful stimuli. *Science* 198:756–758, 1977.

Frederickson, R.C.A.; Nickander, R.; Smithwick, E.L.; Shuman, R.; and Norris, F.H. Pharmacological activity of met-enkephalin and analogues *in vitro* and *in vivo*—depression of single neuronal activity in specified brain regions. In Kosterlitz, H.W. (ed.): *Opiates and Endogenous Opioid Peptides.* Amsterdam: Elsevier, pp. 239–246, 1976.

Garmezy, N. Attentional processes in adult schizophrenia and in children at risk. *J. Psychiatr. Res.* 14:3–34, 1978.

Gilbert, P.E., and Martin, W.R. The effects of morphine and nalorphine-like drugs in the nondependent, morphine-dependent and cyclazocine-dependent chronic spinal dog. *J. Pharmc. Exp. Ther.* 198:66–82, 1976.

Goldstein, A., and Hilgard, E.R. Failure of the opiate antagonist naloxone to modify hypnotic analgesia. *Proc. Natl. Acad. Sci. (USA)* 72:2041–2043, 1975.

Grevert, P., and Goldstein, A. Effects of naloxone on experimental induced ischemic pain and on mood in human subjects. *Proc. Natl. Acad. Sci. (USA)* 74:1291–1294, 1977.

———. Endorphins: Naloxone fails to alter experimental pain or mood in humans. *Science* 199:1093–1095, 1978.

Gunne, L.M.; Lindstrom, L.; and Terenius, L. Naloxone-induced reversal of schizophrenic hallucinations. *J. Neural Transm.* 40:13–19, 1977.

Hall, K.R.L., and Stride, E. The varying response to pain in psychiatric disorders: A study in abnormal psychology. *Br. J. Med. Psychol.* 27: 48–60, 1954.

Hemphill, R.E.; Hall, K.R.L.; and Crookes, T.G. A preliminary report on fatigue and pain tolerance in depressive and psychoneurotic patients. *J. Ment. Sci.* 98:433–440, 1952.

Herman, B.G., and Panksepp, J. Effects of morphine and naloxone on separation distress and approach attachment: Evidence for opiate mediation of social affect.

Hosobuchi, Y.; Adams, J.E.; and Linchitz, R. Pain relief by electrical stimulation of the central gray matter in humans and its reversal by naloxone. *Science* 197:183–186, 1977.

Hosobuchi, Y., and Li., C.H. The analgesic activity of human β-endorphin in man. *Commun. Psychopharmacol.* 2:33–37, 1978a.

———. A demonstration of the analgesic activity of human β-endorphin in six patients. *Pain Abstracts,* vol. 1, Second World Congress on Pain, Montreal, Canada, 1978b.

Hosobuchi, Y.; Rossier, J.; Bloom, F.E.; and Guillemin, R. Stimulation of human periaqueductal gray from pain relief increases immunoreactive β-endorphin in ventricular fluid. *Science* 203:279–281, 1979.

Hughes, J.; Smith, T.W.; Kosterlitz, H.W.; Fothergill, L.A.; Morgan, B.A.; and Morris, H.R. Identification of two related pentapeptides from the brain with potent opiate agonist activity. *Nature* 258:577–580, 1975.

Jacob, J.J.; Tremblay, E.C.; and Colombel, M.D. Facilitation de reactions nociceptives par la naloxone chez la souris et chez le rat. *Psychopharmacologia* 37:217–223, 1974.

Klee, W.A.; Zioudron, C.; and Streaty, R.A. Exorphins: Peptides with opiate activity isolated from wheat gluten and their possible role in the etiology of schizophrenia. In Usdin, E., Bunney, W.E., Jr. and Kline, N.A. (eds.) *Endorphins in Mental Health Research.* London: Macmillan, pp. 209–218, 1979.

von Knorring, L. The experience of pain in patients with depressive disorders. A clinical and experimental study. Umea University Medical Dissertations, Umea, Sweden, 1975.

von Knorring, L.; Almay, B.G.L.; Johansson, F.; and Terenius, L. Pain perception and endorphin levels in cerebrospinal fluid. *Pain* 5:359–365, 1978.

Krieger, D.T.; Liotta, A., and Li, C.H. Human plasma immunoreactive β-lipotropin: Correlation with basal and stimulated ACTH concentrations. *Life Sci.* 21:1771–1778, 1977.

Langley, J.N. On the contraction of muscle, chiefly in relation to the presence of "receptive" substances. Part IV. The effect of curari and some other substances on the nicotine response of the sartorius and gastrocnemius muscles of the frog. *J. Physiol.* 39:253, 1909.

Lasagna, L. Drug interaction in the field of analgesic drugs. *Proc. R. Soc. Med.* 58:978–983, 1965.

Levine, J.D.; Gordon, N.C.; and Fields, H.L. The mechanism of placebo analgesia. *Lancet* 2:654–657, 1978.

————. Naloxone dose dependency produces analgesia and hyperalgesia in postoperative pain. *Nature* 278:740–741, 1979.

Levine, J.D.; Gordon, N.C.; Jones, R.T.; and Fields, H.L. The narcotic antagonist naloxone enhances clinical pain. *Nature* 272:826–827, 1978.

Lindström, L.H.; Widerlöv, E.; Gunne, L.M.; Wahlström, A.; and Terenius, L. Endorphins in human cerebrospinal fluid: Clinical correlations to some psychotic states. *Acta Psychiatr. Scand.* 57:153–164, 1978.

Lord, J.A.H.; Waterfield, A.A.; Hughes, J.; and Kosterlitz, H. Endogenous opioid peptides: multiple antagonists and receptors. *Nature* 267:495–499, 1977.

Mains, R.E.; Eipper, B.A.; and Ling, N. Common precursors to corticotropins and endorphins. *Proc. Natl. Acad. Sci. (USA)* 74:3014–3018, 1977.

Malmo, R.B., and Shagass, C. Physiologic studies of reaction to stress in anxiety and early schizophrenia. *Psychosom. Med.* 11:9–24, 1949.

Malmo, R.B.; Shagass, C.; and Smith, A.A. Responsiveness in chronic schizophrenia. *J. Personality* 19:359–375, 1951.

Marchand, W.E. Occurrence of painless myocardial infarction in psychotic patients. *N. Engl. J. Med.* 253:51–55, 1955.

Marchand, W.E.; Sarota, B.; Marble, H.C.; Leavy, T.M.; and Burbank, C.B. Occurrence of painless acute surgical disorders in psychotic patients. *N. Engl. J. Med.* 260:580–585, 1969.

Martin, W.R.; Eades, C.G.; Thomson, J.A.; Huppler, R.E.; and Gilbert, P.E. The effects of morphine and nalorphine-like drugs in the nondependent and morphine-dependent chronic spinal dog. *J. Pharmac. Exp. Ther.* 197:517–532, 1976.

Mayer, D.J., and Hayes, R.L. Stimulation-produced analgesia: Development of tolerance and cross-tolerance to morphine. *Science* 188:941–943, 1975.

Mayer, D.J., and Liebeskind, J.C. Pain reduction by focal electrical stimulation of the brain: An anatomical and behavioral analysis. *Brain Res.* 68:73–93, 1974.

Mayer, D.J.; Price, D.D.; and Raffi, A. Acupuncture analgesia in man reduced by the narcotic antagonist naloxone. *Brain Res.* 121:368–372, 1977.

Mayer, D.J.; Wolffe, T.L.; Akil, H.; Carder, B.; and Liebeskind, J.C.

Analgesia from electrical stimulation in the brainstem of the rat. *Science* 174:1351–1354, 1971.

Merskey, H. The characteristics of persistent pain in psychological illness. *J. Psychosom. Res.* 9:291–298, 1965*a*.

———. The effect of chronic pain upon the response to noxious stimuli by psychiatric patients. *J. Psychosom. Res.* 8:405–519, 1965*b*.

Mihic, C., and Binkert, E. Is placebo an analgesia mediated by endorphins? Vol. 1, Second World Congress on Pain, Montreal, Canada, 1978.

Oliveras, J.L.; Besson, J.M.; Guilband, G.; and Liebeskind, J.C. Behavioral and electrophysiological evidence of pain inhibition from midbrain stimulation in the cat. *Exp. Brain Res.* 20:32–44, 1974.

Oliveras, J.L.; Hosobuchi, Y.; Redjemi, F.; Guilband, G.; and Besson, J.M. Opiate antagonist, naloxone, strongly reduces analgesia induced by stimulation of a raphe nucleus (centralis inferior). *Brain Res.* 120: 221–229, 1977.

Pert, A.; Pert, C.; Davis, G.C.; and Bunney, W.E., Jr. Opiate peptides and brain function. In van Praag, H. (ed.): *Handbook of Biological Psychiatry,* in press.

Pert, C.B.; Pert, A.; and Tallman, J.F. Isolation of a novel endogenous opiate analgesic from human blood. *Proc. Natl. Acad. Sci. (USA)* 73: 2226–2230, 1976.

Pert, C.B., and Snyder, S.H. Properties of opiate-receptor binding in rat brain. *Proc. Natl. Acad. Sci. (USA)* 70:2243–2247, 1973.

Pomeranz, B., and Chiu, D. Naloxone blockade of acupuncture analgesia: Endorphins implicated. *Life Sci.* 19:1757–1762, 1976.

Reynold, D.V. Surgery in the rat during electrical analgesia induced by focal brain stimulation. *Science* 164:444–445, 1969.

Roberts, J.L., and Herbert, E. Characterization of a common precursor to corticotropin and β-lipotropin: Cell-free synthesis of the precursor and identification of corticotropin peptides in the molecule. *Proc. Natl. Acad. Sci. (USA)* 74:4826–4830, 1977.

Sappington, J. Thresholds of shock-induced discomfort in process and reactive schizophrenics. *Percept. Mot. Skills* 37:489–490, 1973.

Satoh, M.; Kawajiri, S.; Yammoto, M.; Makino, H.; and Takagi, H. Reversal by naloxone of adaptation of rats to noxious stimuli. *Life Sci.* 24:685–690, 1979.

Simon, E.J.; Hiller, J.M.; and Edelman, I. Stereospecific binding of the potent narcotic analgesic [³H]etorphine to rat brain homogenate. *Proc. Natl. Acad. Sci. (USA)* 70:1947–1949, 1973.

Sjölund, B.; Terenius, L.; and Eriksson, M. Increased cerebrospinal fluid levels of endorphins after electro-acupuncture. *Acta Physiol. Scand.* 100:382–384, 1977.

Terenius, L., Characterization of the "receptor" for narcotic analgesics in synaptic plasma membrane fraction from rat brain. *Acta Pharmacol. Toxicol.* 33:377–384, 1973.

Terenius, L., and Wahlström, A. Morphine-like ligand for opiate receptors in human CFS. *Life Sci.* 16:1759–1764, 1975.

Terenius, L.; Wahlström, A.,; Lindström, L.; and Widerlöv, E. Increased CFS levels of endorphins in chronic psychosis. *Neurosci. Letters* 3: 157–162, 1976.

Venables, P.H. Input regulation and psychopathology. In Hammer, M., Salzinger, K., and Sutton, S. (eds.): *Psychopathology.* New York: Wiley, pp. 261–284, 1974.

Watson, S.J.; Berger, P.A.; Akil, H.; Mills, M.J.; and Barchas, J.D. Effects of naloxone on schizophrenia: Reduction in hallucinations in a subpopulation of schizophrenics. *Science* 201:73–76, 1978.

Social and Psychological Factors in Disease

Herbert Weiner

Many and varied categories of social factors have been associated with disease. Conversely, the number of diseases and conditions for which associations with social factors have been reported is very large—essential hypertension, coronary heart and cerebrovascular diseases, cancer, malnutrition, obesity, anorexia nervosa, peptic duodenal ulcer, tuberculosis, rheumatoid arthritis, communicable and metabolic diseases, mental illnesses, alcoholism, and automobile accidents.

A review of all the social factors and their associated diseases is not possible in a short chapter such as this one, which will concern itself with only some, and how they are mediated, adapted to by, or influence the people afflicted. Some of the social factors that promote health or that are associated with disease will be mentioned here, but nothing will be said about those factors that influence or promote the recovery from, or the continuation of, disease, or that affect health services and community well-being. Particular emphasis will be paid to social order and stability as protectors of health; to social and personal customs that promote disease and illness; to social class as a potent factor in the prevalence of certain diseases; and to the increase of specific disease during particular historical periods. An attempt will be made to identify those personal characteristics that lead to adaptive success and failure and thus respectively to health or disease.

The data that support some of the associations between social factors and specific diseases have been criticized for numerous conceptual and methodological reasons. These criticisms are frequently not justified, or are misdirected, for two reasons:

1. Single social factors have been sought to account for the associations. In fact, complex relationships exist—for example, between obesity, social class, ethnicity, and salt intake, in essential hypertension (EH). In some of the best studies, obesity accounts for 20 percent only of blood pressure (BP) variance.
2. Every disease is heterogeneous. At least three forms of early, or so-called borderline hypertension have now been identified that vary in their pathogenesis, pathophysiology, and psychology (Esler et al. 1975, 1977; Julius and Esler 1975). A minimum of two main forms of peptic duodenal ulcer (PDU) and probably four have also been identified (Samloff 1977; Samloff et al. 1975).

57

Thus the need to specify the subforms of a disease becomes urgent in the search for its social and psychological correlates. I review here these associations.

Social Stability and Order

Henry and Cassel (1969) have summarized abundant data that contravene the cherished belief that BP levels increase with age. Cultural and subcultural groups exist that show no increase of BP with age; these groups have been identified in widely separated geographic locations, and diverse ethnic origins. Henry and Cassel's review lead them to conclude that (1) living in a stable society in which social order, tradition, and traditional roles are maintained, and rule-following behavior is prescribed; and (2) allowing an individual to deal with a predictable environment are associated with no rise in BP in the individual members of such groups, regardless of whether they are U.S. naval aviators, or some persons in India.

A similar conclusion was arrived at by those who have studied coronary heart disease. This conclusion is not surprising, because high BP is a major risk factor in coronary heart disease (CHD) (Kannel 1977). Although the results of the Framingham study are that multiple risk factors—for example, smoking, glucose intolerance, raised cholesterol levels, the type A personality, and a short stature—are cumulative in CHD, the fact remains that the level of BP is the prepotent risk factor: the higher the level the greater the risk. Or to put it another way, increases in BP contribute to CHD in proportion to the degree of BP elevation; they do so independently of the other risk factors enumerated (Kannel 1977).

A longitudinal study on the mortality of myocardial infarction was carried out by Stout, Bruhn, Wolf, and others in 1964 and 1966, in Roseto, Pennsylvania, a community of southern Italian immigrants who have retained their traditional patriarchal ways, and who have trusting and cohesive relationships. Mutually supportive, the rules and the roles of the members of this town are well defined. Their diet contains 41 percent fat. The members of Roseto are an average of 20 pounds overweight. By contrast in an adjacent town, Bangor, the population is of mixed English, German, and Italian stock. Families there are less gregarious and religious, and their roles are less clearly defined. The death rate from myocardial infarction (MI) in Roseto was one-half of that in Bangor. By 1976, however, as the younger inhabitants of Roseto were leaving and entering the mainstream of American life, the incidence of MI in the younger group is increasing.

Studies contrasting Japanese and Japanese-Americans, Benedictine monks working in the community and their cloistered brothers in the United States, and nomadic Bedouins and Bedouin settlers in villages in Israel come to similar conclusions (Caffrey 1966; Marmot 1975; Marmot and Syme 1976; Matsumoto 1970).

Nonetheless the problem of how to interpret such data remains. Is it affluence, diet, smoking, social mobility, or incongruity that accounts for these differences, or combinations of these? Should the formulation that social order and stability *protect* against both hypertension and CHD be true, one would predict that social disruption and lack of order are conducive to increased BP levels. Indeed in a seminal report, Harburg and his coworkers (1973) found such a positive relationship. They reported that in Detroit poverty, social and marital disruption, skin color, police brutality, and a high crime rate were positively correlated with BP levels. The absence of social order is a major factor producing elevated BP levels when other factors such as obesity and salt intake are taken into account.

Conversely, where social order and stability prevail—when the rate and amount of social change is slow and slight and when social supports are available—the adaptive tasks confronting persons are tolerable and minimal; ambiguity is reduced; and the social milieu is predictable.

Historical and Political Conditions

The incidence and prevalence of most diseases rises and falls. Historical periods when external factors of a political or other nature create extraordinary conditions—for instance, war—are associated with remarkable changes in the incidence of some diseases. For 300 years it has been known that during wartime starvation induces epidemic amenorrhea, the amenorrhea of starvation (LeRoy Ladurie 1969), although it was not until World War I that this form of amenorrhea was first studied by physicians. During World War II there was a twofold increase of another disease, proved duodenal ulcer (DU), in the U.S. army. Following the war the incidence and prevalence of DU fell to prewar levels. A marked increase in perforated DUs occurred in London, England, but only during the period of the German bombing raids.

The effects of war, and of military occupation of a country by a brutal conqueror, were associated with the fivefold increase in the incidence of thyrotoxicosis—both the toxic nodular and the toxic diffuse varieties—in Denmark and Norway during the Nazi occupation (Grelland 1946; Iversen 1948, 1949; Meulengracht 1949). However, in occupied Holland from 1940 to 1945 a *decrease* of thyrotoxicosis occurred (Schweitzer 1944). Why? The reasons given are that from the very beginning of their occupation the Germans attempted to starve the Dutch into submission; not so the Danes. In 1943 the Germans reduced the availability of food to the Norwegians to marginal levels (1,570 KCal per day). Subsequently, the incidence of thyrotoxicosis in Norway fell. But no such regimen of starvation was prescribed to the Danes, and the incidence remained high until 1946. The students of this phenomenon ascribe it to the interaction of military occupation, danger, an increase in infectious disease, and nutrition.

Socioeconomic Status

Low socioeconomic (SE) status in many communities has been associated with shorter life expectancy, the increased risk for mental disorders, higher infant mortality, cervical cancer in women, social and nutritional deprivation in infants and children, and obesity. Obesity in turn is a hazard predisposing to diabetes mellitus, contributing to high BP, cholelithiasis, gout, osteoarthritis, the Pickwickian syndrome, the sleep-apnea syndrome, and carcinoma of the breast and endometrium (DeWaard 1975; Mirra et al. 1971; Wynder 1966).

Obesity is unevenly distributed in the various social classes; it decreases in prevalence with increasing SE status. But social class of origin is almost as closely linked to obesity as is the subject's own social class. Obviously, a person's own social class cannot influence his parents'. Hence the relationship must be unidirectional (Goldblatt, Moore, and Stunkard 1965).

Curiously enough, the observed relationship with social class obtains in primary anorexia nervosa (AN). The observation has been made repeatedly that AN occurs in those girls of elevated SE status. Crisp, Palmer, and Kalucy (1976) found that in an English private girls' school the prevalence of AN was 1:100 in 16–18-year-old girls, while in a state school it was 1:550. The association between SE status and AN is valid but has resisted understanding; the factors that mediate this relationship are unknown.

Social and Personal Customs

Social customs may determine who does or does not develop a disease. And personal habits such as smoking and drug abuse may place persons at risk for a variety of diseases.

In some tribes in New Guinea, ritual cannibalism has been associated with a deadly viral disease, kuru (Gadjusek 1976). Because the preparation of the infected brains of the deceased relatives is largely carried out by women, and the prepared brains are eaten mainly by women and children, this relentlessly progressive and eventually fatal disease is three times as prevalent in them than in adult men. Therefore, the sex ratio in these groups is determined by the disease, and until recently men predominated in villages in which kuru was indigenous. With the abolition of cannibalism, kuru is becoming less and less prevalent.

Smoking cigarettes is a risk factor in a variety of diseases, of which the most important are carcinoma of the lung, coronary heart disease, and peptic ulcer. The increasing incidence of carcinoma of the lung in women has been directly related to their increased use of cigarettes since the first quarter of this century.

Occupation

The list of occupational and industrial hazards is unfortunately a long one—exposure to asbestos, coal and cement dusts, chemicals, and ionizing radiation is well-known to place persons at risk for malignancies, leukemias, and other pathologies.

However, certain occupations seem to be particularly stressful not only chemically but also psychologically. Cobb and Rose (1973) have documented an increased incidence of high blood pressure, duodenal ulcer, and probably diabetes mellitus in men who control air traffic—an occupation that requires responsibility, precision, alertness, and attentiveness. Note that this stressful occupation is not associated with only one disease.

Technological Advances

Technological advances create new occupations that are in turn stressful. Such advances and discoveries create new environments in which man works, lives, and plays; but they may also alter his body's internal environment.

New drugs are potentially hazardous. The use of the oral contraceptive pill (OCP) has undoubtedly led to the resurgence of venereal diseases. But it also places women at risk for thromboembolic disease, high BP, and carcinoma of the cervix. Curiously enough, Fries and Nillius (1973) have found that 5 percent of a series of young women who stopped taking the OCP later developed frank anorexia nervosa.

Social Change

We come now to a more dynamic series of studies that deal with the consequences of social and political changes, which may either promote health or protect against disease. Kotchen et al. (1974) have reported a study of BP levels in adolescents following school desegregation in the United States. They found that 10 percent of black youths, 1 percent of black girls, and 0 percent of white youths and girls had systolic BP levels above 140 mm Hg when studied in two schools, one of which was integrated and the other of which was not. Of black youths, 8.5 percent in the integrated school in a middle-class neighborhood, and 12.3 percent of black youths in the ghetto school had elevated BP levels. Among the blacks significantly higher blood pressure levels were found in children whose parents worked as laborers, or who were unemployed, than in children who had parents working in professions.

Kotchen and his collaborators partialed out the effects of weight. Weight alone accounted for only 20 percent of the BP variance. When adjusted for differences in weight, inner-city adolescents still had higher BP levels than whites. Clearly ethnicity, socioeconomic background, and weight all play a role in adolescent high BP. (No single factor accounts for increased BP levels).

All disease is multifactorial. Furthermore, disease cannot be understood if one does not also take into account the multiple vulnerability of persons. This generalization is buried in all the data presented so far because (1) not every black adolescent in a ghetto school develops high BP; (2) not every woman taking the OCP develops the same illness, and some develop none; and (3) not every air traffic controller develops the same disease, and some develop none.

All these studies also deal with associations between some social or environmental variables and important diseases. They are essentially "structural" studies rather than transactional ones: they do not deal with three significant questions:

1. Why does one person fall ill with one disease and not another?
2. Why does the incidence or prevalence of most disease change with time?
3. Why does the predisposed person fall ill at a particular time in his or her life and not another?

The answer to the first question is that multiple social, psychological, and physiological predispositions exist for most diseases. The answer to the second question is largely unknown. The answer to the third question is approached by clinical and more systematic studies that deal with environmental change: for example, Engel's studies (1968) on bereavement as a general onset condition of many diseases, and Holmes and Rahe's studies (1967) of life change. Despite many reservations, Holmes and Rahe (1967) contains an important grain of truth that confirms what every wise physician can observe every day. Yet its association of life changes and illness onset is too linear. It fails to take into account other variables such as the perception of the life change, as distressing or not, as a critical variable in the association (Lundberg, Theorell, and Lund 1975), as well as the successful or unsuccessful psychological adaptation to the change; and the presence or absence of social support as playing crucial, intervening roles in modifying the effects of life changes (DeFaire and Theorell 1976).

We are now more on psychological territory. Clearly, life changes alone do not determine disease onset. As Hinkle (1974) showed, healthy persons frequently tend to be contented in their personal lives and occupations. Contentment can insulate persons from change. Other data suggest that social change is not directly pathogenetic: it does not directly act upon the individual. The change must be perceived and appraised, activities that in turn are influenced by past experience, role, and status relationships (Lazarus 1966).

To summarize the pathogenetic role of life experience:

1. Life experiences may be overwhelming, for example, living in London during the Blitz; fighting in Stalingrad; or being exposed to explosions. The former experience led to increased perforation of the duodenum, the two latter to high BP.
2. The meaning of the experience may idiosyncratically be perceived as portentous, or the experience may be perceived as excessively dangerous, leading to a "paralysis" of appropriate coping measures.
3. The significance of the experience may be denied or inappropriately responded to. In other persons the signals of distress or danger fail, or they are not as perceived (Nemiah and Sifneos 1970).
4. The person may be vulnerable because the event mobilizes specific, unmastered, personal problems with their own antecedents in that person's history.
5. Some persons (children, the defective, the dependent, the widowed, the elderly) have limited adaptive capacities, or do not have the education, skills, social supports, or information to cope with the events and to solve the personal problems the events create.

Many psychological and psychiatric studies of medically ill patients have described their poor adaptive and coping capacities, dependency, and primitive defensive measures that result in adaptive failure; that are signaled by helplessness and hopelessness; and that finally lead to giving-up (Engel 1968; Ruesch 1948; Vaillant 1977; Weiner 1977). Conversely, effective coping is determined by a combination of psychological skills—intelligence, the capacity to use information and solve problems, familiarity with the task, self-reliance and self-confidence, independence, realism, hopefulness in the face of challenge. But the task is made easier when the environment is relatively stable and predictable.

Summary

The relationships and associations between social and psychological factors and disease are complex and nonlinear. At first glance, this conclusion might be the product of the fact that our understanding of most of the diseases discussed in this review is far from complete. However, this conclusion is not warranted. In the cases of viral and parasitic diseases, about which our knowledge is much more profound, a similar complex, nonlinear model obtains (Bloom 1979; Blumberg 1977). Such a model can be summarized as follows (Weiner 1978).

1. A pool of persons predisposed to a disease occurs in every population.
2. The time of life a person is infected and the route of infection determine in part the course and nature of the disease or carrier state.
3. The same inciting agent can give rise to multiple disease forms, which are determined by the characteristic of the agent in interaction with the host.
4. The adaptation (immune response) to the agent can vary quantitatively or qualitatively.
5. Varying adaptive responses determine in part the ultimate form in which the disease is expressed.
6. Social and cultural factors play a role in disease and are in turn influenced by it.
7. The disease can be transmitted horizontally and vertically in several ways.

We can make some generalizations on the basis of these conclusions. For example, the factors that predispose to a disease are not invariantly those that incite or sustain it (Weiner 1977). In addition, in every disease genetic polymorphism plays a complex role. The age and sex of individuals determines their responses to disease. The phenotypic expression of disease is a complex product of the interaction of the agent and the host in which qualitative and quantitative variables in the adaptive-defensive responses of the host play significant roles.

References

Bloom, B.R. 1979. Games parasites play: How parasites evade human surveillance. *Nature* (London) 279:21-26.

Blumberg, B.S. 1977. Australia antigen and the biology of hepatitis B. *Science* 197:17-25.

Bruhn, J.G.; Chandler, B.; Miller, M.C.; et al. 1966. Social aspects of coronary heart disease in two adjacent, ethnically different communities. *Am. J. Public Health* 56:1493-1506.

Caffrey, C.B. 1966. Behavior patterns and personality characteristics as related to prevalence rates of coronary heart disease in Trappist and Benedictine monks. Ph.D. dissertation (Clinical Psychology), Catholic

University of America, Washington, D.C. (University Microfilms, Inc., Ann Arbor, Michigan, no. 67-1830, pp. 45-48 coronary heart disease rates).

Cobb, S., and Rose, R.M. 1973. Hypertension, peptic ulcer, and diabetes in air traffic controllers. *J.A.M.A.* 224-489.

Crisp, A.H.; Palmer, R.L.; and Kalucy, R.S. 1976. How common is Anorexia Nervosa? A prevalence study. *Br. J. Psychiat.* 128-549.

De Faire, U., and Theorell, T. 1976. Life changes and myocardial infarction. How useful are life change measurements? *Scand. J. Soc. Med.* 4:115-122.

DeWaard, F. 1975. Breast cancer incidence and nutritional status with particular reference to body weight and height. *Cancer Res.* 35:3351-3356.

Engel, G.L. 1968. A life setting conducive to illness: The giving-up, given-up complex. *Arch. Intern. Med.* 69:293-300.

Esler, M.D.; Julius, S.; Randall, O.S.; et al. 1975. Relation of renin status to neurotic vascular resistance in borderline hypertension. *Amer. J. Cardiol.* 36-708.

Esler, M.D.; Julius, S.; Zweifler, A.; et al. 1977. Mild high-renin essential hypertension. *N. Engl. J. Med.* 296-405.

Fries, H., and Nillius, S.J. 1973. Dieting, anorexia nervosa and amenorrhea after oral contraceptive treatment. *Acta Psychiat. Scand.* 49:669-679.

Gadjusek, D.C. 1976. Unconventional viruses and the origin and disappearance of kuru. In *Les Prix Nobel en 1976.* Stockholm: P.A. Norstedt and Soner, 1976.

Goldblatt, P.B.; Moore, M.E.; and Stunkard, A.J. 1965. Social factors in obesity. *J.A.M.A.* 192:1039-1044.

Grelland, R. 1946. Thyrotoxicosis at Ulleval Hospital in the years 1934-1944 with a special view to frequency of the disease. *Acta Med. Scand.* 125:108.

Harburg, E.; Erfurt, J.C.; Hauenstein, L.S.; et al. 1973. Socio-ecological stress, suppressed hostility, skin color, and black-white male blood pressure: Detroit. *Psychosom. Med.* 35:276.

Henry, J.P.; and Cassel, J.C. 1969. Psychosocial factors in essential hypertension: Recent epidemiologic and animal experimental evidence. *Am. J. Epidemiol.* 90:171-200.

Hinkle, L.E., Jr. 1974. The effect of exposure to culture change, social change, and changes in interpersonal relationships on health. In *Stressful Life Events: Their Nature and Effects,* B.S. Dohrenwend and B.P. Dohrenwend, eds., pp. 9-45. New York: Wiley.

Holmes, T.H., and Rahe, R.H. 1967. The social readjustment rating scale. *J. Psychosom. Res.* 11:213-218.

Iversen, K. 1948. *Temporary Rise in the Frequency of Thyrotoxicosis in Denmark 1941-1945.* Copenhagen: Rosenskilde and Bagger.

————. An epidemic wave of thyrotoxicosis in Denmark during World War II. *Amer. J. Med. Sci.* 217:121.

Julius, S., and Esler, M. 1975. Autonomic nervous cardiovascular regulation in borderline hypertension. *Amer. J. Cardiol.* 36:685.

Kannel, W.B. 1977. Importance of hypertension as a major risk in cardiovascular disease. In *Hypertension,* J. Genest, E. Koiw, and O. Kuchel, eds. New York: McGraw-Hill.

Kotchen, J.M.; Kotchen, T.A.; Schwertman, N.C.; et al. 1974. Blood pressure distributions of urban adolescents. *Amer. J. Epidemiol.* 99:315.

Lazarus, R.S. 1966. *Psychological Stress and the Coping Process.* New York: McGraw-Hill.

LeRoy Ladurie, E. 1969. Famine amenorrhea (seventeenth–twentieth centuries). *Annales E.S.C.* 24:1589–1601.

Lundberg, U.; Theorell, T.; and Lind, E. 1975. Life changes and myocardial infarction: Individual differences in life change scaling. *J. Psychosom. Res.* 19:27.

Marmot, M.G. 1975. Acculturation and coronary heart disease in Japanese-Americans. Ph.D. dissertation (Epidemiology), University of California, Berkeley.

Marmot, M.G., and Syme, S.L. 1976. Acculturation and coronary heart disease in Japanese-Americans. *Am. J. Epidemiol.* 104:225–247.

Matsumoto, Y.S. 1970. Social stress and coronary heart disease in Japan. A hypothesis. *Milbank Mem. Fund Q.* 48:9–36.

Meulengracht, E. 1949. Epidemiologic aspects of thyrotoxicosis. *Arch. Intern. Med.* 83:119.

Mirra, A.P.; Cole, P.; and MacMahon, B. 1971. Breast cancer in an area of high parity: Sao Paulo, Brazil. *Cancer Res.* 31:77–83.

Nemiah, J.C., and Sifneos, P.E. 1970. Affect and fantasy in patients with psychosomatic disorders. In *Modern Trends in Psychosomatic Medicine,* vol. 2, O.W. Hill, ed. London: Butterworths.

Ruesch, J. 1948. The infantile personality: The core problem of psychosomatic medicine. *Psychosom. Med.* 10:134.

Samloff, I.M. 1977. Radioimmunoassay of group II pepsinogens in serum. *Gastroenterology* 72:1125.

Samloff, I.M.; Liebman, W.M.; and Panitch, M. 1975. Serum group I pepsinogens by radioimmunoassay in control subjects and patients with peptic ulcer. *Gastroenterology* 69:83–90.

Schweitzer, P.M.J. 1944. Calorie supply and basal metabolism. *Acta Med. Scand.* 119–306.

Stout, C.; Morrow, J.; Brandt, E.N., Jr.; et al. 1964. Unusually low incidence of death from myocardial infarction. Study of an Italian-American community in Pennsylvania. *J. Amer. Med. Assoc.* 188:845–849.

Vaillant, G.E. 1977. *Adaptation to Life.* Boston: Little, Brown.

Weiner, H. 1977. *Psychobiology and Human Disease.* Elsevier North-Holland, Inc.

Weiner, H. 1978. The illusion of simplicity: The medical model revisited. *Am. J. Psychiat.* 135:Suppl. 27–33.

Wynder, E.L.; Escher, G.C.; and Mantel, N.A. 1966. An epidemiological investigation of cancer of the endometrium. *Cancer* 19:489–520.

6

An Analysis of Causes of Sex Differences in Mortality and Morbidity

Ingrid Waldron

In 1978 in the United States, the life expectancy for females was 77.2 years, while the life expectancy for males was only 69.5 years (U.S. NCHS 1980*a*). This difference of 7.7 years in life expectancy reflects substantially higher death rates for males than for females.

In contrast to the higher death rates for men, for many measures of ill health or morbidity rates are higher for women. This is illustrated in figure 6–1 for the 45–64-year-old age group in the United States. Data from the national Health Interview Survey indicate higher rates for women for acute conditions, for days of restricted activity or bed rest due to illness, and for doctor visits (U.S. NCHS 1974*a*, 1977*a,b,* 1978*a,b,* 1979*b,* 1980*b*). On the other hand, men are reported to have somewhat higher rates of chronic conditions that limit activities. For the 45–64-year-old age group, data from the Health Interview Survey indicate similar numbers of hospitalizations for men and women, but longer hospitalizations for men, resulting in more days of hospitalization per person for men. For younger adults, higher rates of hospitalization are reported for women, even after exclusion of maternity hospitalizations (U.S. NCHS 1976*a*).[1] Also, for younger adults the excess of doctor visits by women is substantially higher than at older ages, even after exclusion of visits associated with pregnancy (Nathanson 1978). With these exceptions, the sex differences in morbidity and mortality illustrated in figure 6–1 are generally representative of sex differences reported for adults in the United States in the last decade (U.S. NCHS 1974*a,* 1976*a,* 1977*a,b,* 1978*a,b,* 1979*b,* 1980*a,b;* Nathanson, 1977*a,* 1978; Verbrugge, 1976 *a,b*).

These observations lead directly to several major questions. Why do men have higher mortality and a shorter life expectancy than women? Why do women report more acute conditions, more disability days due to illness, and more visits to physicians than men? Why is there such a considerable variation in sex differences, ranging from a substantial male excess for mortality, through a small male excess for chronic conditions reported to result in activity limitation, to a female excess for reported acute conditions, dis-

I am grateful to Belinda Wagner for her assistance in gathering data used in this chapter and to Teresa Sullivan for helpful comments on an earlier draft of this chapter.

69

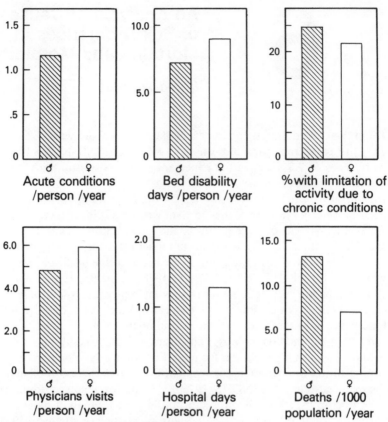

Source: Data from U.S. NCHS 1978*a*, 1979*a*.

All morbidity data were derived from the national Health Interview Survey.

Note: Acute conditions are conditions that lasted less than three months and that involved medical attention or restricted activity, excluding conditions on a specified list of potentially chronic conditions such as allergy, arthritis, or hypertension. For acute conditions only, the data in the figure include all persons aged 45 and over; the incidence of reported acute conditions decreases with age, so the rates for 45–64-year-olds would be somewhat higher than those shown in the figure. A day of bed disability is a day on which a person stayed in bed for over half the day because of illness or injury. Physician visits include any consultation with a physician, except for patients in hospitals. Hospital days are for short-stay hospitals only.

Figure 6–1. Sex Differences in Reported Morbidity and Mortality for 45–64-Year-Olds, United States, 1977

ability days, and doctor visits. These questions have been addressed by previous authors who have offered a variety of explanations for these phenomena. The major explanatory hypotheses will be briefly reviewed in this introductory section. These hypotheses will be evaluated in subsequent sections in the context of an examination of the evidence concerning sex differences in health. Although these hypotheses have sometimes been presented as mutually exclusive, they will be treated here as potentially complementary explanations of the reported sex differences. The evidence to be presented indicates that the reported sex differences reflect the effects of multiple, interacting factors.

Madigan (1957) and others have hypothesized that the higher mortality of males is due primarily to genetic differences that result in an inherently greater vulnerability of males. Postulated inherent advantages of females include protection by female sex hormones against the risk of coronary heart disease and inherently greater immune resistance for females (Waldron 1976). Other authors have emphasized the importance of behavioral and cultural factors. Retherford (1975) has argued that men's higher death rates are due in large part to men's higher rates of cigarette smoking. Waldron (1976, 1980a), Nathanson (1977a, b), and Harrison (1978) have argued for the importance of a broader range of behaviors associated with the male role, including the hard-driving Type A or coronary-prone behavior pattern, employment in hazardous occupations, and alcohol consumption, in addition to cigarette smoking.

In analyzing the sex differences in morbidity, Mechanic (1976), Nathanson (1977a, 1978) and Verbrugge (1976a) have emphasized that sex differences in reporting of illness and in behavior in response to illness may have a substantial impact on many of the available measures of morbidity. These authors have hypothesized that women may be more aware of symptoms and illness and may also be more willing to report symptoms or illness in an interview. These authors have also hypothesized that, for a given health condition, women may be more likely to visit a doctor and to reduce activity or to take a day of rest. It may be socially more acceptable for women to be concerned with their health, and it may be easier for women who are housewives to visit doctors or to reduce activity, because the housewife role permits more flexible use of time than employment does. These hypotheses suggest that behavioral differences between men and women may have a major impact on the reported sex differences in morbidity. From this perspective, women's higher rates of self-reported acute conditions, disability days, and doctor visits may not reflect actual differences in the physical morbidity experienced by men and women. Gove and Hughes (1979), on the other hand, have argued that women do have higher rates of minor illness than men, and that this is due primarily to women's poorer mental health and to role obligations that interfere with women's ability to care for their health.

Why is there such great variation in the sex differences for different health measures? If sex differences in the various morbidity measures are differently influenced by sex differences in reporting and behavior in response to illness, then this could account for much of the variation in sex differences for different health measures (Nathanson 1977*a;* Verbrugge 1976*a*). Sex differences in behavior in response to illness may even contribute to the sex differences in mortality. Women may have lower death rates in part because they get more rest when ill and get more medical care (Waldron 1976).

Additional hypotheses will be presented in subsequent sections of this chapter. The explanatory hypotheses will be evaluated in the context of an analysis of data concerning sex differences in mortality and morbidity. Causes of the sex differences in mortality in the contemporary United States are analyzed in the next section of this chapter, and causes of sex differences in morbidity are analyzed in the following section. The last section of the chapter briefly reviews evidence concerning cross-cultural and historical variation in sex differences in mortality.

Sex Differences in Mortality in the Contemporary United States

Table 6–1 gives the sex-mortality ratio for each major cause of death in the United States in 1978. The sex mortality ratio, that is, the ratio of male to female death rates, is highest for seven causes of death with major behavioral components. These are homicide, suicide, motor vehicle accidents, other accidents, lung cancer and emphysema, which are both closely linked to cigarette smoking (U.S. HEW, Surgeon General 1979), and cirrhosis of the liver, which is related to excessive alcohol consumption (Galambos 1979). Taken together, these causes of death are responsible for 35 percent of the excess of male over female death rates. The sex mortality ratio for ischemic heart disease is somewhat lower, but because it is such a major cause of death, ischemic heart disease by itself is responsible for 38 percent of the excess of male over female death rates. The following paragraphs summarize evidence concerning the causes of the sex difference for each of these causes of death. The evidence presented indicates that behaviors that have been more expected or acceptable for males make a major contribution to the higher mortality of males.

Malignant Neoplasms of the Respiratory System and Emphysema

Men's death rates are much higher than women's death rates for malignant neoplasms of the respiratory system, primarily lung cancer, and for emphysema, bronchitis, and asthma, primarily emphysema (table 6–1). More men than women smoke cigarettes and, among those who smoke, men smoke more cigarettes per day, men inhale more, and men are less likely to use filter or low-tar cigarettes (U.S. HEW, Surgeon General 1979; Hammond

Table 6-1
Sex Mortality Ratios for All Major Causes of Death, United States, 1978

Cause of Death	Sex Mortality Ratio[a]	Male Death Rate[b]	Female Death Rate[b]
Homicide	3.64	15.3	4.2
Malignant neoplasms of respiratory system	3.53	59.0	16.7
Suicide	2.98	18.5	6.2
Emphysema, bronchitis, and asthma	2.92	11.1	3.8
Motor vehicle accidents	2.85	35.1	12.3
All other accidents	2.85	31.4	11.0
Cirrhosis of the liver	2.17	17.6	8.1
Ischemic heart disease	2.13	256.3	120.6
Influenza and pneumonia	1.83	21.0	11.5
Other malignant neoplasms	1.72	49.9	29.0
Other cardiovascular diseases	1.69	46.6	27.6
All other causes of death	1.63	105.2	64.6
Malignant neoplasms of digestive organs and peritoneum	1.55	41.6	26.9
Certain causes of mortality in early infancy	1.29	11.9	9.2
Arteriosclerosis	1.28	6.8	5.3
Cerebrovascular diseases	1.19	49.8	41.8
Malignant neoplasms of genital organs	1.04	15.2	14.6
Diabetes mellitus	1.02	10.5	10.3
Malignant neoplasms of breast	0.01	0.2	23.1
All causes	1.80	802.8	447.0

Source: Data from U.S. NCHS (1980a).

[a]The sex mortality ratio is the ratio of male to female death rates.

[b]Death rates are deaths/100,000 population. These death rates have been age adjusted using the age distribution of the 1940 U.S. population as the standard. Age adjustment ensures that male and female death rates are directly comparable and are not affected by the higher proportion of females at older ages. Causes responsible for less than 1 percent of deaths have been grouped in residual categories.

1966). These sex differences in smoking habits are a major cause of men's elevated rates of lung cancer and emphysema (U.S. HEW, Surgeon General 1979). The sex mortality ratios for lung cancer and emphysema have been found to be much lower for nonsmokers than for the total population including smokers (table 6-2).

Why do more men than women smoke? Smoking by women was strongly discouraged by the social mores of the early twentieth century. The

Table 6-2
Comparison of Sex Mortality Ratios for Nonsmokers and for Total Population

	Sex Mortality Ratios	
Cause of Death	For Those Who Never Smoked Regularly	For the Total Sample
Lung cancer		
Ages 45–64	1.6	7.3
65–79	1.4	9.4
Emphysema		
Ages 45–64	4.0	11.7
65–79	2.2	7.3
Coronary heart disease		
Ages 45–54	4.5	7.5
55–64	3.3	4.4
65–74	2.1	2.4
All causes of death		
Ages 45–54	1.3	2.2
55–64	1.7	2.5
65–74	1.6	2.0

Source: Adapted from Waldron (1976); data from Hammond (1966).

Figures for lung cancer and emphysema are approximate due to incomplete published data.

conventions of that period continue to influence the smoking patterns of people who were teenagers at that time, since relatively few people begin smoking cigarettes after age 20. As a consequence, there is a particularly large sex differential in cigarette smoking for older people who were teenagers before 1930 and thus were over 65 in 1975 (U.S. HEW, Surgeon General 1979). Another factor that may have contributed to sex differences in cigarette smoking is the link between rebelliousness and teenage cigarette smoking (U.S. HEW, NCI 1977; U.S. HEW, Surgeon General 1979). In general, girls have tended to be less rebellious and more conforming to adult standards, probably in part because parents and teachers of school-age children have allowed boys more independence and expected girls to be more obedient (Maccoby and Jacklin 1974; Waldron 1976). Girls' lesser rebelliousness may be one reason why, until very recently, teenage girls have been less likely than teenage boys to begin smoking cigarettes (U.S. HEW, NCI 1977; U.S. HEW, Surgeon General 1979). An additional factor that may have contributed to sex differences in cigarette smoking may be a greater susceptibility of girls than boys to nicotine-overdose reactions (Silverstein et al. 1980). It appears that girls more often than boys feel sick as a result of smoking their first cigarette, and this difference may have contributed to girls' lower rates of cigarette smoking in the period before low-nicotine cigarettes were in widespread use.

Although the sex differences in lung cancer and emphysema rates have been smaller for nonsmokers than for the total population, nevertheless, even among nonsmokers, men have had higher lung cancer and emphysema death rates than women (table 6-2). In addition, men who were current smokers have also had higher death rates for these causes than women who were current smokers, and the increase in risk for cigarette smokers has been found to be greater for men than for women (U.S. HEW, Surgeon General 1979). There are at least two important reasons for these sex differences. First, men's smoking habits result in a greater risk of lung cancer and emphysema, since male smokers smoke more cigarettes per day, inhale more, are less likely to use filter or low-tar cigarettes, and, until recently, have also begun smoking at younger ages (U.S. HEW, Surgeon General 1979; Hammond 1966). When men and women are matched for tar consumption over a fifteen-year period or for approximate total number of cigarettes ever smoked, then the proportionate increase in risk of lung cancer appears to be at least as high for women as for men (Mushinski and Stellman 1978; Williams and Horn 1977).

The second important reason for sex differences in the risk of lung cancer and other lung diseases is that men are more often exposed to occupational hazards that increase their risk of lung diseases (U.S. DOL 1980). One major hazard is asbestos, which is widely used in insulation and construction materials. Asbestos insulation workers have been found to have an increase in risk of lung cancer ranging from approximately two-fold to eight-fold, depending on characteristics such as the latency from initial exposure (Selikoff and Lee 1978; Selikoff et al. 1979; Selikoff et al. 1980). Asbestos exposure appears to result in both an increased risk of lung cancer for nonsmokers and a larger increase of risk for cigarette smokers (Selikoff et al. 1980; Hammond et al. 1979). One and a half million workers in the United States are currently exposed to asbestos on the job (NIOSH 1977), and Bridbord et al. (1978) have estimated that four million workers in the United States have been heavily exposed to asbestos in past or present jobs. Since most of the jobs involving heavy asbestos exposure have been held by men, this widespread exposure to a potent carcinogen makes a significant contribution to the sex difference in lung cancer mortality. Based on levels of risk and exposure, it is estimated that roughly one-tenth of men's lung cancer deaths in the United States are related to asbestos exposure.

Exposure to petroleum products, especially polynuclear aromatic hydrocarbons, is also associated with a significantly increased risk of lung cancer (Menck and Henderson 1976). Almost four million workers in the United States are exposed, including workers in the petrochemical industry, vehicle drivers, roofers, and metal workers (NIOSH 1977; Menck and Henderson 1976). Other occupational carcinogens, such as metallic dusts and fumes, also contribute to increased risk of lung cancer for men (Bridbord et al. 1978; Waldron 1976). Although data on exposure and risks are incomplete, the available data are sufficient to suggest that occupational exposures play a role in a substantial fraction of men's lung cancer deaths.

Ischemic or Coronary Heart Disease

A recent revision of the International Classification of Diseases has resulted in the official designation of *ischemic heart disease* for the category commonly known as coronary heart disease (Havlik and Feinleib 1979). Because the term *coronary heart disease* is more familiar and is used in most of the studies cited, this term will be used in the text of this chapter. The major forms of coronary heart disease include myocardial infarctions, which are commonly known as heart attacks, and angina pectoris.

Death rates for coronary heart disease are twice as high for men as for women in the contemporary United States (table 6-1). Because coronary heart disease is such a major cause of death, coronary heart disease by itself is responsible for 38 percent of the total sex difference in mortality. The sex difference in risk of coronary heart disease is linked to sex differences in the extent of atherosclerosis of the coronary arteries (figure 6-2) (Strong et al. 1978; McGill and Stern 1979). (Atherosclerotic plaques in the coronary arteries can decrease the flow of blood to the heart muscle, and inadequate blood flow is the primary proximal cause of coronary heart disease.) Elevated blood pressure and serum cholesterol are major risk factors for the development of coronary heart disease, but sex differences in these risk factors are relatively small and do not appear to be the primary cause of the sex differences in coronary heart disease (figure 6-2) (Waldron 1980a; McGill and Stern 1979).

A major cause of the sex differences in mortality due to coronary heart disease are the sex differences in cigarette smoking habits which have been described above. Cigarette smoking is associated with increased atherosclerosis of the coronary arteries (U.S. HEW, Surgeon General 1979), and the sex differences in cigarette smoking presumably contribute to the observed sex differences in coronary atherosclerosis.

When comparisons were made between men and women who had never smoked regularly, sex differences in coronary heart disease were substantially smaller than the sex differences for the total population, including cigarette smokers (table 6-2). Depending on the age group, approximately

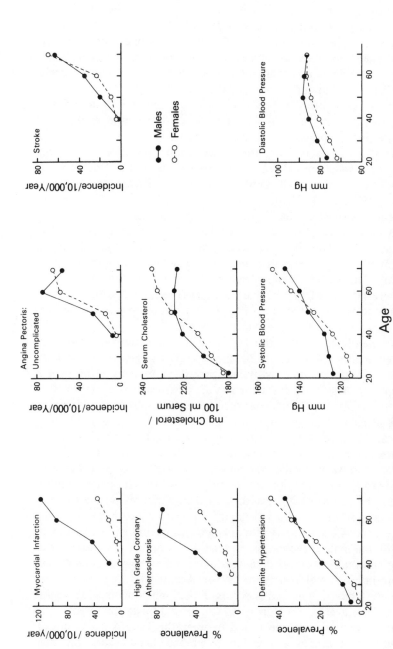

Figure 6-2. Sex Differences in Cardiovascular Morbidity in the United States

Source: Reprinted with permission from I. Waldron, 1980a, Sex differences in longevity. In S.G. Haynes and M. Feinleib (eds.), *Second Conference on Epidemiology of Aging*, p. 167.

Note: Incidence data are from the Framingham study (Shurtleff 1970). Data on atherosclerosis are from a sample from Minnesota (Stamler et al. 1970). Cholesterol and blood pressure data are for U.S. national samples (U.S. NCHS 1976b, 1977c).

a quarter to three-fifths of the sex difference in coronary heart disease was related to cigarette smoking (Waldron 1980*a*). Thus, cigarette smoking appears to be a major cause of sex differences in coronary heart disease. However, the substantial sex differences in coronary heart disease mortality for nonsmokers indicate that there must be other factors that also contribute to sex differences in coronary heart disease.

One factor that appears to play a major role is the Type A or coronary prone behavior pattern, a hard-driving style of life characterized by a chronic sense of time urgency, a strong drive to achieve, competitiveness, aggressiveness, hostility, and impatience (Rosenman et al. 1976; Jenkins 1976). Men or women with this behavior pattern have a substantially increased risk of developing or dying of coronary heart disease (Jenkins 1976; Haynes et al. 1978, 1980; Waldron 1978*a*).

As might be expected, the Type A behavior pattern is more common among men than among women in the United States (Waldron 1978*a;* Haynes et al. 1978). This sex difference in behavior pattern probably contributes to men's higher rates of coronary heart disease mortality (Waldron 1976, 1980*a*). The Type A behavior pattern appears to contribute to increased atherosclerosis of the coronary arteries (Frank et al. 1978), and the sex difference in behavior pattern probably contributes to the observed sex difference in coronary atherosclerosis (figure 6-2).

Why do men have more of the Type A behavior pattern than women? This difference does not appear to be due primarily to genetic differences. Twin studies have not found a significant genetic component of the Type A behavior pattern (Matthews and Kranz 1976; Rosenman et al. 1976). Furthermore, this behavior pattern appears to be uncommon for men as well as for women in many other cultures (Cohen 1974; Dentan 1968). This evidence indicates that the sex differences in Type A behavior pattern in the United States are probably due in large part to cultural factors.

There are two ways in which the male role in the United States appears to contribute to greater development of the Type A behavior pattern among males. First, the competitive pressures and time pressures on the job may foster this hard-driving, hurried behavior pattern (Waldron et al. 1980). Second, the Type A behavior pattern appears to contribute more to success in the traditional male role than in the traditional female role, and consequently the development of this behavior pattern may be encouraged more in boys than in girls (Waldron 1976, 1978*a*). Adults who have the Type A behavior pattern have higher status occupations, higher incomes, and more education on the average (Waldron 1978*a*). In contrast, for middle-aged women, no association has been found between the Type A behavior pattern and whether a woman is married or whether her husband has high socioeconomic status (Waldron 1978*b*). Similarly, for college students, the Type A behavior pattern is associated with better grades, but it is not associated with success in relationships with the opposite sex (Waldron et al. 1980). Thus, the Type A behavior pattern may be encouraged more in boys than in girls because this behavior pattern appears to contribute to success in the vocational sphere, which has traditionally been emphasized for men, but it does not appear to contribute to success in at least one goal that has traditionally been emphasized for women, namely, "marrying well."

There has been considerable interest in the possible contribution of sex hormones to the sex differences in coronary heart disease. It has been hypothesized that estrogens reduce women's risk of coronary heart disease. Administration of estrogens generally reduces levels of serum cholesterol and low-density lipoproteins, factors that are associated with increased risk of coronary heart disease (McGill and Stern 1979; Waldron 1976). Also, administration of estrogens generally increases the levels of high-density lipoproteins, which are associated with a reduced risk of coronary heart disease. In addition, a number of studies have found that both natural menopause and surgical menopause were associated with an increased risk of coronary heart disease, although other studies have not found this effect (Waldron 1976; Gordon et al. 1978; McGill and Stern 1979). It is unclear to what extent the association between surgical menopause and increased risk of coronary heart disease should be interpreted as evidence of a protective effect of women's hormones; since women who have had only their uteri removed appear to have normal hormone levels and yet they also have an increased risk of coronary heart disease (Waldron 1976; Gordon et al. 1978).

Studies of the effects of estrogen therapy on the risk of coronary heart disease indicate that exogenous estrogens do not, in general, reduce the risk of coronary heart disease. Most studies of estrogen therapy in men report an increase in risk of coronary heart disease (Waldron 1976; McGill and Stern 1979). Various studies have reported that estrogen therapy after menopause is associated with increased risk, decreased risk, or no change in the risk of coronary heart disease (Gordon et al. 1978; Waldron 1976; McGill and Stern 1979). Use of oral contraceptives appears to be associated with an increased risk of coronary heart disease, and the estrogen component of oral contraceptives appears to be responsible for the increase in risk (McGill and Stern 1979). The increase in risk associated with estrogen therapy in men or with oral contraceptive use in women appears to be related to the fact that estrogen enhances blood coagulation processes, and this can lead to decreased blood flow in the coronary arteries (Waldron 1976; McGill and Stern 1979).

In summary, estrogens have diverse and counteracting effects on the physiological processes involved in the pathogenesis of coronary heart disease. Exogenous estrogens do not in general reduce the risk of coronary heart disease. Evidence concerning the effects of endogenous estrogens is contradictory, and it remains to be determined whether endogenous estrogens reduce the risk of coronary heart disease for women.

Progestogens, the other type of female sex hormone, do not appear to have a substantial effect on the risk of coronary heart disease (Waldron 1976; McGill and Stern 1979). Male sex hormones do not appear to elevate the risk of coronary heart disease (Waldron 1978c, Nordøy et al. 1979). For example, castration of men is apparently not associated with significant changes in the progression of coronary atherosclerosis. Other biological differences between the sexes, such as differences in the thickness of the intima of the coronary arteries, have been proposed as factors responsible for sex differences in the risk of coronary heart disease, but none of these hypotheses is supported by convincing evidence at the present time (McGill and Stern 1979).

In conclusion, sex differences in coronary heart disease mortality appear to be due in large part to sex differences in cigarette smoking and the Type A behavior pattern. Biological differences, such as possible protective effects of estrogens, may contribute to the sex differences in coronary heart disease, but evidence concerning this is contradictory. As will be discussed in the final section of this chapter, the sex differences in coronary heart disease mortality are much smaller in some other countries than in the United States (figure 6-3) (Stamler et al. 1970). This observation is compatible with the conclusion that cigarette smoking and the Type A behavior pattern make major contributions to the sex difference in coronary heart disease mortality, since both of these behaviors are strongly influenced by cultural factors and are much less common in many other societies than in the United States. On the other hand, there does not appear to be any country where women have higher coronary heart disease mortality than men, and this suggests that biological differences between the sexes tend to protect women and/or that sex differences in cigarette smoking and Type A behavior have been influenced by similar factors cross-culturally so that in no country have rates been higher for women than for men.

Accidents and Cirrhosis

Males in the United States have almost three times as many fatal motor vehicle accidents as females (table 6-1). This is due in part to the fact that men drive more, but a major cause appears to be that men drive less safely. Male drivers are involved in approximately twice as many fatal accidents per mile driven as female drivers (National Safety Council 1980). One study has found that men were more likely to enter an intersection after the light had turned yellow or red, and more likely to fail to signal a turn (Waldron 1976). Men may also drive more at times when roads are crowded or under

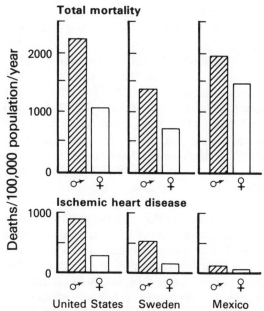

Source: Data from World Health Organization, 1975.

Figure 6-3. Total Mortality and Ischemic Heart Disease Mortality for 55-64 Year Olds in the United States, Sweden and Mexico, 1971 and 1972

hazardous conditions, and this may also contribute to their higher accident rates.

For accidents other than motor vehicle accidents, death rates for males are almost three times higher than rates for females (table 6-1). About one-third of the male excess is due to accidents while at work.[2] The rate of work accidents is higher for men than for women because more men are employed and because the jobs men hold are more physically hazardous on the average (Waldron 1980b). An additional 40 percent of the male excess for fatal nonmotor vehicle accidents is due to gun accidents, accidental drownings, and air and water transport accidents, which are each over five times more common for males (data from U.S. NCHS 1979c). It is evident from these statistics that men's higher accident fatalities are due in large part to behaviors which are encouraged in boys and men: driving cars, working at sometimes hazardous jobs, using guns, being adventurous, and acting unafraid. The expectation that boys and men will be more adventurous and take more risks than girls and women is conveyed to children in many ways, including the stories written for children (Waldron 1976).

Another cause of men's higher accident fatalities is their greater consumption of alcohol (Blum 1967; Cahalan 1970). Half of all fatal motor vehicle accidents involve drunken drivers. Other accidents are also associated with alcohol use. Surveys of alcohol levels in drivers have found that a higher proportion of men than of women drivers have high alcohol levels (Ferrence 1980). For drivers involved in fatal motor vehicle accidents, men more often than women were under the influence of alcohol. Several studies of drinking drivers who were responsible for a fatal accident have found that a large proportion were alcoholics or at least had serious drinking problems (Fine and Scoles 1974). Rates of alcoholism, problem drinking, or heavy drinking are several times higher for men than for women (Ferrence 1980; Cahalan 1970). To the extent that drunken drivers themselves are victims of the fatal accidents they cause, men's higher rates of heavy drinking make a significant contribution to their higher motor vehicle accident death rates.

Excessive alcohol consumption and the malnutrition that is a frequent concomitant greatly increase the risk of cirrhosis of the liver (Galambos 1979; Lelbach 1974). The greater prevalence of heavy drinking among men appears to be the primary reason why men's death rates for cirrhosis of the liver are twice as high as women's (table 6–1) (Galambos 1979). Women appear to have at least as high a susceptibility as men to cirrhosis of the liver due to infectious liver disease. For a given level of alcohol consumption, women appear to be as susceptible or more susceptible than men to alcohol-induced liver injury (Galambos 1979). Thus, it appears that the major reason why men have higher rates of cirrhosis of the liver is that more men are heavy drinkers.

Why do more men than women drink alcohol and drink heavily? Cruz-Coke (1971) has argued that an X-linked recessive gene may be responsible for men's higher rates of alcoholism, but evidence presented by Winokur (1967) raises serious doubts concerning the validity of this hypothesis. Furthermore, Cloninger and co-workers (1978) have presented persuasive evidence that sex differences in alcoholism are due not to genetic factors but rather to environmental factors. The importance of cultural influences is indicated by the observation that in one-third of nonindustrial societies studied, both sexes consumed alcohol equally (Bacon 1973). Bacon has noted that equal alcohol consumption for men and women has been particularly common in societies where alcohol has been introduced recently, and she hypothesizes that social proscriptions against heavy drinking by women develop in time, because even occasional inebriation may have disastrous consequences for the care of babies and small children, while occasional drunkenness may be more compatible with the usual responsibilities of men in nonindustrial societies. Studies in the contemporary United States have shown that there is greater acceptance of drinking by men, and this may contribute to the higher rates of heavy drinking among men (Cahalan 1970; Lawrence and Maxwell 1962).

Homicide and Suicide

Men are more than three times as likely as women to be homicide victims, and five times as likely as women to be arrested as suspected murderers (table 6–1). (U.S. FBI 1964, 1972, 1980). Approximately a quarter of murders involve family members or lovers (U.S. FBI 1972, 1974, 1980). For this category, the number of females involved nearly equals the number of males, both as victims and as murderers (U.S. FBI 1972, 1974; Wolfgang 1958). Approximately a third of homicides involve "other arguments" (U.S. FBI 1972, 1974, 1980). "The vast majority in this category were the result of impulsive rage involving a wide range of altercations, such as arguments over a cigarette, ice cream, noise, etc." (U.S. FBI 1964). "The persons participating in these arguments were most frequently acquainted prior to the fatal act" (U.S. FBI 1972). In this category, men are much more likely than women to be involved as victims and as perpetrators (U.S. FBI 1964; Wolfgang 1958). Men also outnumber women in the other major category, murders related to another felony, such as robbery.

Higher alcohol consumption by men probably contributes to men's higher homicide rates (Wolfgang 1958; Blum 1967). About half of homicide perpetrators and victims have been drinking prior to the murder (Blum 1964). A high incidence of alcohol consumption among homicide victims is of special significance in light of the finding that homicides are frequently victim precipitated in the sense that the victim was the first to strike a blow or to show a deadly weapon. In Wolfgang's sample, 31 percent of homicides with male victims were victim precipitated, whereas only 6 percent of homicides with female victims were victim precipitated.

Suicide rates are three times as high for men as for women, but women make twice as many suicide attempts as men (table 6–1) (Jarvis et al. 1976; Schneidman and Farberow 1961). Suicidal men are particularly likely to use guns, with consequences which are irreversible and fatal, whereas women are more likely to use poisons, which can be treated by the use of stomach pumps and antidotes (Kramer et al. 1972). However, for every method there are more male than female suicides, so some additional explanation must be sought for the preponderance of male suicides and female suicide attempts.

A variety of evidence indicates that a suicide attempt is frequently a desperate, last-ditch plea for help, rather than an actual attempt to end life (Stengel and Cook 1958). Women apparently are better able to use a suicide attempt as a plea for help, and it seems likely that this ability to some extent protects them from the need to actually kill themselves (Peck and Schrut 1971). In contrast, males "see themselves as strong, powerful, dominant, potent" (Maccoby and Jacklin 1974) and find it difficult to seek help. This is probably one reason why men are more likely to carry a suicidal act through to a fatal conclusion.

Jarvis et al. (1976) have postulated that individuals who are more socially integrated are more likely to make a suicide attempt rather than to commit suicide. They argue that women are more socially integrated because women marry earlier and as young adults more often live with family or friends. This may be a reason why the sex mortality ratios for suicide are high at ages 15–24 (U.S. NCHS 1979c). However, it is doubtful whether sex differences in social integration can account for sex differences in suicidal behavior at older ages. Sex mortality ratios for suicide increase progressively from ages 40 to 85, and the highest sex mortality ratios are observed at older ages (U.S. NCHS 1979c); yet, at older ages men are much more likely than women to be married and women are much more likely to be living alone (U.S. Bureau of the Census, 1975, 1977).

Other Causes of Death

The male excess of deaths for influenza and pneumonia and other infectious diseases may be due in part to genetically determined sex differences in immune resistance (Waldron 1976; Purtilo and Sullivan 1979). For example, the X chromosome appears to carry quantitative genes for the production of immunoglobin M which result in higher serum levels of immunoglobin M in females. Infectious diseases are responsible for only about 4 percent of the total sex difference in mortality in the contemporary United States (Waldron 1980c). Lower immune resistance of males may also contribute to higher rates of malignant neoplasms.

Sex differences in reproductive anatomy and physiology contribute to sex differences in cancer mortality. For example, death rates for malignant neoplasms of the breast are about 100 times higher for women than for men (table 6–1). The normal stimulatory action of estrogens on breast growth seems to extend to stimulation of the development of breast cancers; removal of the ovaries before age 35 results in a decrease of over two-thirds in breast cancer rates at older ages (McMahon et al. 1973).

Total cancer mortality is higher for males than for females, but if cancers related to cigarette smoking are excluded, sex differences in cancer mortality are very small (Waldron 1976, 1980c). Correspondingly, for a large sample of nonsmokers, no sex differences in total cancer mortality were observed (Hammond and Seidman 1980). This suggests that sex differences in cigarette smoking may be the most important contributor to sex differences in total cancer mortality, and sex differences in risk due to differences in immune responses, reproductive physiology, exposure to occupational carcinogens, and other factors may be approximately balanced for the two sexes.

Conclusion

This analysis has shown that, in the contemporary United States, men have higher mortality than women in large part because of behaviors which have been expected of or been more socially acceptable for males. Behaviors which are common among men and contribute to men's higher mortality include cigarette smoking, excessive consumption of alcohol, employment in jobs that involve exposure to carcinogens or risk of accidents, the hard-driving Type A behavior pattern, and use of guns. Genetically determined physiological differences between males and females may also contribute to sex differences in mortality, although this type of effect appears to be much less important than the behavioral differences. A possible protective effect of endogenous estrogens against coronary heart disease may be the most important genetically determined physiological difference, although the evidence concerning this is contradictory.

Sex Differences in Morbidity in the Contemporary United States

As outlined at the beginning of this chapter, the substantial male excess observed for mortality is not observed for most measures of morbidity. Data from the Health Interview Survey indicate somewhat higher rates of activity-limiting chronic illness for men, but higher rates for women for acute illness, doctor visits, and days of restricted activity or bed rest due to illness (figure 6-1) (U.S. NCHS 1974*a*, 1977*a,b*, 1978*a,b*, 1979*b*, 1980*b*). This section begins with an analysis of sex differences in chronic illness, since the pattern of sex differences in this case is most similar to the pattern of sex differences in mortality described in the previous section. The discussion of chronic conditions is followed by an analysis of the reasons for women's higher rates for many morbidity measures.

Chronic Conditions

The national Health Interview Survey provides information concerning rates of reported chronic conditions that resulted in restricted activity and that lasted for at least three months, or that were one of a specified list of potentially chronic conditions such as asthma or hypertension (U.S. NCHS 1979*b*). Higher rates are reported for men for chronic conditions that limit a person's major activity or that cause any activity limitation (figure 6-1) (U.S. NCHS 1974*a*, 1978*a*, 1979*b;* Verbrugge, 1976*a,b*). However, rates were higher for women for reports of having any chronic condition

(Verbrugge 1976a). (This last statement is based on surveys prior to 1968, since more recent data are not available.) These data may be strongly influenced by factors other than the actual prevalence of physical morbidity. For example, as discussed below, women may be more likely to be aware of and to report a given physical symptom or illness. On the other hand, one reason why activity limitation is reported less frequently for women than for men may be that women may be better able to adjust to a given physical condition without perceiving a limitation of their activity because women's roles may be more flexible and less physically strenuous on the average (Verbrugge 1976a).

Table 6-3, which is adapted from the work of Verbrugge (1976a), shows the sex morbidity ratios and sex mortality ratios for a number of major causes of chronic illness and mortality. It can be seen that, for most causes, the sex morbidity ratio for activity-limiting chronic conditions is similar to the sex mortality ratio. This suggests that the sex differences in self-reported activity-limiting chronic conditions are due in large part to factors similar to those responsible for the sex differences in mortality.

Studies of medically diagnosed disease provide further evidence that for many specific causes the sex differences in chronic morbidity and mortality are similar. For example, for major types of cancer there is a close correspondence between sex mortality ratios and sex morbidity ratios based on age-adjusted incidence of medically diagnosed cancer from a major survey covering 10 percent of the U.S. population in 1969–1971 (Cutler et al. 1974). For whites, the sex morbidity ratios were 1.5 for cancers of the digestive system and peritoneum, 1.1 for cancers of the genital organs, and .01 for breast cancers. The sex morbidity ratios for blacks were somewhat higher in each case. These sex morbidity ratios correspond closely to the sex mortality ratios for these types of cancer given in table 6-1. For cancers of the respiratory system sex ratios have declined substantially in recent years, so a closer time correpondence is required for comparisons of morbidity and mortality data. For 1970 the sex mortality ratio for cancers of the respiratory system was 5.2 (data from U.S. NCHS 1974c), which corresponds closely to the sex morbidity ratio of 5.0 for whites and a somewhat higher ratio for blacks. For stroke or cerebrovascular disease, studies of medically diagnosed morbidity (figure 6-2) (Epstein et al. 1965) and mortality statistics (table 6-1) both show slightly higher rates for men than for women.

Although for many chronic conditions the sex ratios for morbidity and mortality are quite similar, in a few cases there are substantial discrepancies. The most striking example is hypertension. In national surveys, higher rates are reported for women both for "high blood pressure" (U.S. NCHS 1976b) and for hypertension resulting in limitation of major activity (table 6-3). In contrast, there is a male excess for mortality due to hypertension or hypertensive heart disease (table 6-3) (U.S. NCHS 1980a). Studies of the

Table 6–3

Sex Mortality and Sex Morbidity Ratios for Leading Causes of Death or Morbidity, United States, 1958–1972

Cause of Death or Illness	Sex Mortality Ratio	Sex Morbidity Ratio	Female Death Rate	Female Prevalence of Morbidity
Emphysema		4.8		0.9%
Asthma	4.3	1.4	4.7	.42
Bronchitis		1.2		.08
Injuries	2.8	1.5	41.7	a
Heart conditions	1.9	1.3	182.6	1.33
Malignant neoplasms	1.4	1.0	109.1	.17
Hypertension without heart disease	1.3	.53	2.7	.55
Cerebrovascular disease	1.2	1.6	63.3	.19
Diabetes mellitus	.86	.83	15.1	.42
Arthritis and rheumatism	.68	.68	1.0	1.69

Sources: Verbrugge, 1976a; U.S. NCHS 1974b.

Note: Causes were included if it was possible to make a "perfect" or "near perfect" match between morbidity and mortality codes and if the cause was responsible for more than 1% of all deaths or had a prevalence of more than .5%. Both the sex mortality ratios and the sex morbidity ratios give ratios of male to female rates. With the exception of the data for injuries, the morbidity data are for prevalence of chronic conditions reported to result in limitation of a person's major activity. Sex ratios are for 1958–1972. Morbidity prevalence data are for 1969–1970. Death rates are per 100,000 population for 1970 or 1969. Both death rates and morbidity rates are age-adjusted, with the total 1940 U.S. population as the standard.

[a]Incidence of acute conditions for injuries = 33/100 people/year.

prevalence of hypertension or hypertensive heart disease based on clinical examination and blood pressure measurements for national samples have yielded results generally similar to the mortality data and have not shown the large excess for women observed in the survey interview data (U.S. NCHS 1974d, 1978d). Why then do women report more hypertension? One probable explanation is that women are more likely than men to be aware of their hypertension, since women make more doctor visits than men (U.S. NCHS 1978b, 1979b, 1980b), and hypertension is frequently not symptomatic and yet is easily detected at any doctor visit. Community surveys of blood pressure have found that a higher proportion of women than men had been aware of their high blood pressure prior to the survey (U.S. NCHS 1978d).

The sex mortality ratio for heart conditions was 1.9, while the sex morbidity ratio was only 1.3 for heart conditions reported to limit a person's major activity (table 6–3). For clinically diagnosed morbidity, rates of myocardial infarction have been over twice as high for men as for women, while sex differences for angina pectoris have been much smaller (figure 6-2; U.S. NCHS 1965). Myocardial infarctions are associated with a greater mortality risk than is angina pectoris (Kannel et al. 1979; Weinblatt et al. 1973). The data indicate that the difference between the sex mortality and sex morbidity ratios reflects in large part a greater contribution of myocardial infarctions to mortality rates and a greater contribution of angina pectoris to morbidity rates. Interestingly, although the mortality risk associated with a myocardial infarction is similar for men and women, the mortality associated with angina is lower for women than for men (Kannel et al. 1979; Kannel and Feinleib 1972; Weinblatt et al. 1973). This suggests that the angina identified in women is less serious than the angina identified in men. In addition, angina is more likely to be first diagnosed as a sequel of myocardial infarction for men, whereas angina is more often the presenting complaint for women (Kannel and Feinleib 1972). These observations suggest that women may have a greater tendency to perceive and to report symptoms that lead to a diagnosis of angina, and this may be one reason why sex differences are smaller for angina than for myocardial infarctions.

The sex mortality ratio for injuries is also much higher than the corresponding sex morbidity ratio (table 6-3). Several behavioral differences between the sexes appear to play a role. For example, as described in the preceding section, women are more likely to make a suicide attempt and less likely to actually commit suicide, in part because they are less likely to use guns and probably also because of sex differences in the willingness to use a suicide attempt as a plea for help. Male automobile drivers have a much larger excess for fatal accidents per mile driven than for nonfatal accidents (National Safety Council 1980), and this probably reflects sex differences in the type of driving, driving speeds, and frequency of driving while intoxicated.

In summary, for many specific causes, the sex ratios for activity-limiting chronic illness are similar to the sex ratios for mortality. This suggests that similar causal factors probably contribute to the sex differences in morbidity and in mortality. For some causes there are quantitative differences or even reversals between the sex differences for morbidity and mortality. A variety of behavioral factors, such as frequency of doctor visits and use of guns, appear to contribute to the discrepancies in these cases.

Finally, it should be noted that while total male mortality exceeded female mortality by 80 percent in the United States in 1978 (table 6-1), the prevalence of chronic conditions reported to result in limitation of any activity or limitation of major activity was only 11–16 percent higher for males (based on age-adjusted rates calculated from data in U.S. NCHS 1979b). This difference is due in part to higher sex mortality than sex morbidity ratios for some specific causes, and to reversals of sex differences in a few cases. In addition, as Verbrugge (1976a) has noted, certain causes with a female excess make a much larger contribution to total morbidity than to total mortality, while other causes with a substantial male excess make a much larger contribution to total mortality (table 6–3).

Acute Illness, Doctor Visits, and Disability Days

As discussed at the beginning of this chapter, women report higher rates of acute conditions, more days of restricted activity due to illness, and more days of bed rest due to illness, and women make more physician visits. To some extent these sex differences are the result of women's more complex and demanding reproductive functions. Visits for prenatal care alone account for about one-quarter of the sex differences in doctor visits at ages 15–44 (data from U.S. NCHS 1980b). One study in the United States (Chien and Schneiderman 1975) and another in Denmark (Hollnagel and Kamper-Jørgensen 1980) indicate that for middle-aged adults about half of the sex difference in doctor visits is due to pregnancy and sex-specific conditions. Data from the Health Interview Survey indicate that pregnancy and genito-urinary disorders also make a major contribution to women's higher rates of acute conditions (U.S. NCHS 1979b).

Although reproduction-related disorders make a substantial contribution to women's higher rates of doctor visits and acute conditions, it is clear that other factors also play an important role. For acute conditions, the other categories that make the largest contribution to women's higher rates are respiratory conditions and infective and parasitic diseases (U.S. NCHS 1979b). Women's higher rates for respiratory conditions and for infective and parasitic diseases appear paradoxical for two reasons. First, women have lower rates for the corresponding causes of death (Verbrugge 1976a). Second, genetic differences between the sexes appear to result in somewhat greater immune resistance for women (Waldron 1976; Purtilo and Sullivan 1979). Why then do women report higher rates of acute conditions for these causes? I will consider this question together with the related questions, Why do women visit doctors more often than men, even for nonreproduction-related causes? and Why do women report more disability days than men?

Mechanic (1976), Verbrugge (1976a), and Nathanson (1978) have proposed that women's higher rates of reported acute conditions, disability days, and physician visits may be due in large part to sex differences in reporting illness and in behavior in response to illness. If, for a given level of physical illness, women are more likely than men to report an illness, to visit a doctor, or to take a day of rest, then this could account for the paradox that women report higher rates of acute conditions, doctor visits, and disability days, but men have higher death rates.

Evidence that there are sex differences in the recognition and reporting of illness comes from studies that have compared self-reports of illness with the results of clinical examinations. Several studies indicate that women are more likely than men to report illness not found clinically, and men appear to be somewhat more likely to underreport illness (U.S. NCHS 1967, 1973a; Maddox 1964; Leo et al. 1957; Hollnagel and Kamper-Jørgensen 1980). In addition, in the national Health Interview Survey, proxy reporting by women for men in their household is common, and proxy reporting appears to result in underreporting of illness (U.S. NCHS 1973b).

Higher rates of perceived symptoms for women probably contribute to women's higher rates of doctor visits. In addition, for a given degree of self-reported illness, women appear to make more doctor visits than men (Taylor et al. 1975; Tessler et al. 1976), although sex differences were not entirely consistent in one major study (Wolinsky 1978). In addition, women more frequently have a medical examination for preventive purposes, that is, when not seeking treatment for any illness (Nathanson 1977b; U.S. NCHS 1980c). On the other hand, women delay as long as men before seeking medical attention after the first symptoms of cancer (Hackett et al. 1973; Worden and Weisman 1975) or of myocardial infarction (Simon et al. 1972; Hackett and Cassem 1969; Moss et al. 1969).

There is some evidence that even in childhood there may be sex differences in the propensity to seek medical help. When children were allowed to decide for themselves whether to visit a school clinic, girls made substantially more visits than boys (Lewis et al. 1977). In contrast, national surveys of children's physician visits (which are probably determined primarily by parents' decisions) show approximately equal rates for boys and girls under 17 (U.S. NCHS 1979b). This contrast suggests that on their own initiative girls may be more likely than boys to seek medical attention, even though parents perceive girls and boys as requiring similar levels of medical attention.

There has been some speculation that women make more use of medical services in part because they are encouraged or pressured to do so by physicians. One area for which data are suggestive is gynecological surgery. Rates of hysterectomy have been substantially higher in the United States than in England, and this corresponds to the much higher number of surgeons per capita in the United States (Bunker 1970). Much of the gynecological surgery in the United States appears to be unnecessary, as indicated by the fact that rates have been reduced substantially following the initiation of a medical audit or a requirement for consultation (Bunker 1970). Verbrugge (1978) has reported that during a doctor visit women received more services and more often were given an appointment for a return visit, but men were more often referred to another physician. Armitage et al. (1979) found that for a number of specific complaints men received more extensive workups than women. McCranie et al. (1978) found no evidence of sex bias in physicians' responses to case descriptions. There appears to be little or no sex difference in the proportion of patients who comply with doctor's recommendations (Marston 1970). Thus, on the basis of current evidence it is not clear what contribution, if any, physicians' behavior may make to the sex differences in rates of physician visits.

It has been postulated that it is easier for women to visit a doctor or to restrict activity or to take a day of bed rest, because the housewife role is more flexible than employment and about half of adult women in the United States are housewives (U.S. Bureau of the Census 1975). In the case of doctor visits, the available evidence does not support this hypothesis. Two recent national surveys concerning problems in obtaining medical care have found very small, inconsistent sex differences in the proportion who reported a problem due to inconvenient office hours, doctor not available when needed, or not having enough time (U.S. NCHS 1978e, 1980c). Overall, men were less likely than women to report that they had had a problem in obtaining medical care.

Although women consistently report higher rates of restricted activity days and bed disability days, one study (Gove and Hughes 1979) has found that women are more likely than men to report that when they are really sick they are unable to get a good rest and they have a number of chores that they must continue to do. A probable interpretation is that men less often restrict their activity or spend more than half the day in bed, in part due to constraints imposed by employment, but once men do decide to take a day of rest, they are less likely than women to have to carry on with child care, meal preparation, or similar chores. Thus, men probably experience different types of interference with their ability to rest than housewives do. Employed women probably experience both types of problems in getting rest when ill.

The relative importance of the different types of restriction on rest when ill has not been established. There is, however, evidence that rest days and associated behavior and attitudes may be of considerable importance. One study has shown that adults who reported that they did not take any days of rest due to illness during a one-year period had higher mortality during the subsequent five years than adults who reported that they took one to three days of rest (Berkman 1975). More men than women reported zero days of rest, and it is possible that this difference contributes to men's higher mortality. On the other hand, Gove and Hughes (1979) have argued that women's inability to get a good rest when ill contributes to their higher rates of minor illness. They found that for adults who lived alone there was very little sex difference in reported ability to get rest when ill or in other "nurturant role demands," and also little difference in an index of psychiatric symptoms. Correspondingly, for adults who lived alone there were no significant sex differences in reported health problems during a two-week period or in self-ratings of health, after controlling for income and other demographic variables.[3]

I turn now to several related questions. Do women experience more stress than men? Do women and men respond to stress differently on the average? Do sex differences in stress or in responses to stress contribute to the reported sex differences in morbidity? Stress is a complex concept that is difficult to define or to measure, and at least at present, there does not appear to be a satisfactory measure that could be used to determine definitely whether women or men experience more stress. Recent studies have not reported consistent sex differences in levels of life stress (Uhlenhuth et al. 1974; Mellinger et al. 1978). It does, however, seem clear that on the average men and women experience somewhat different types of stress and that they tend to respond to stress in different ways. For example, for a given level of life stress, women appear to be more likely to report symptoms, to visit a physician, or to take psychotherapeutic medication, while men appear to be more likely to drink alcohol (Otto 1979; Mellinger et al. 1978). It seems likely that males and females are socialized to respond to stress differently and that these differences contribute, on the one hand, to women's higher rates of self-reported acute conditions and doctor visits and, on the other hand, to men's higher rates of injuries and liver disease.

In summary, women's higher rates of doctor visits and self-reported acute conditions appear to be related to behavioral factors such as sex differences in the recognition and reporting of illness and sex differences in typical patterns of response to stress. In addition, women's more complex and demanding reproductive functions contribute to their higher rates of doctor visits and acute conditions.

Marital Status

Are sex differences in health related to marital status? Table 6–4 presents data from national morbidity surveys that show that the pattern of sex differences by marital status varies for different morbidity measures. For acute conditions and activity-limiting chronic conditions, the sex morbidity ratios were lowest for the formerly married. This reflects a particularly large disadvantage for women who were divorced, separated, or widowed, perhaps due to the low income of many women in that group (U.S. Bureau of the Census 1975). For residency in nursing homes, the sex morbidity ratio was lowest for those who were married, reflecting a particularly large advantage associated with marriage for men. A similar pattern, with a greater advantage associated with marriage for men, has been observed for total mortality, suicide, homicide, accident, cirrhosis of the liver, lung cancer, tuberculosis, and diabetes mortality (Gove 1973). It may be that marriage plays

Table 6–4

Relation of Marital Status to Sex Differences in Morbidity, United States, 1971–1972

	Sex Morbidity Ratios		
	Married	Never Married	Formerly Married
Incidence of acute conditions	.84	.85	.76
Prevalence of activity-limiting chronic conditions	1.27	1.33	1.16
Restricted activity days	.86	.99	.90
Physician visits	.67	.72	.68
Residents in nursing homes	.73	.92	1.10

Source: Data from U.S. NCHS 1976e.

Note: Sex morbidity ratios shown are the ratio of the male to female age-adjusted morbidity rates for persons over 16 years old. All data are from the national Health Interview Survey. Data for physician's visits include pre- and postpartum visits. Data for hospital admissions are not shown since deliveries are a major cause of hospital admissions for married women and data excluding deliveries were not available.

a more important beneficial role for men in terms of care received, social support, and improved psychological well-being (Gove 1973; Bock and Webber 1972).

Employment

The foregoing analyses have suggested that employment may make several important contributions to sex differences in morbidity and mortality. Time constraints imposed by employment may contribute to reduced rates of restricted activity days and bed disability days for men. Exposure to occupational carcinogens contributes to men's higher lung-cancer rates, and accidents on the job are a major cause of higher non-motor-vehicle accident rates for men. Men's higher rates of coronary heart disease are related to the greater prevalence among men of the Type A behavior pattern, which may be encouraged more among males because it appears to contribute more to occupational success than to success in traditional female roles.

These observations suggest the hypothesis that sex differences in morbidity and mortality might be smaller for comparisons between men and employed women than for comparisons between men and housewives. Data for restricted activity and bed disability days conform to this hypothesis, with rates for employed women intermediate between the high rates for all

Table 6–5
Relationships of Women's Employment to Sex Differences in Morbidity and Mortality, United States

	All Men	All Women	Employed Women	Housewives
Restricted activity days[a]	13.0	18.1	14.4	NA
Bed disability days[a]	4.1	7.2	5.8	NA
Incidence of acute conditions[b]	197	246	250	NA
Prevalence of activity-limiting chonic conditions[c]	9.2%	8.6%	NA	10.6%
Mortality[d]	2.7	1.5	1.1	NA

Source: Data from Decoufle (1977).

[a]Disability days per person per year, 1975, ages 25–44 years (U.S. NCHS 1978f). These data and the data for acute and chronic conditions are from the national Health Interview Survey.

[b]Acute conditions resulting in restricted activity or medical attention per 100 persons per year, 1977–1978, ages 17–44 years (U.S. NCHS 1979d).

[c]1974, ages 17–44 years (U.S. NCHS 1977d).

[d]Probability of dying during next 8 years, multiplied by 100, from U.S. mortality data for 1968 and Social Security records, 1965–1972; averaged for ages 21–45 years.

women and the low rates for men (table 6–5). However, for both chronic conditions and mortality, available data indicate that employed women are healthier than housewives, and the differences between employed women and men are even greater than the differences between housewives and men (table 6–5).[4]

One important reason why death rates and rates of self-reported chronic illness are lower for employed women than for housewives is that women who are in poor health are less likely to seek employment and are more likely to leave the labor force (Waldron 1980b; Waldron et al. 1981). Thus, in comparisons of the health of employed women and housewives, the effects of employment on health are obscured by the effects of health on women's employment.

Another reason why employed women do not have higher chronic illness and mortality rates comparable to men's rates may be that, on the average, employment appears to have a more harmful effect on the health of men than on the health of women (Waldron 1980b). Men are more often employed in hazardous jobs with greater exposure to occupational carcinogens or to risk of serious accidents. In addition, a major effect of employment on men's health may be indirect—the effect that expectations of the male role has on the socialization of behaviors such as the Type A behavior pattern. This type of effect is presumably much more widespread for men than for employed women in the United States, at least at present.

Finally, two general explanations can be offered for the differences between employed women and men in morbidity and mortality. First, sex differences in morbidity and mortality appear to be strongly influenced by differences in health-related behaviors, and these are probably influenced more by differences in socialization during childhood than by adult experience in the labor force. Second, there are substantial differences in the roles and responsibilities of employed women and men. For example, employed women usually carry primary responsibility for care of family and home and thus have less time available for rest (Robinson et al. 1972). These role differences may also contribute to the reported morbidity differences between employed women and men.

Cross-Cultural and Historical Variation in Sex Differences in Mortality

This section reviews evidence concerning stability or variability of sex differences in mortality in different countries and in different historical periods. Findings of consistent sex differences are suggestive of an influence of genetically determined differences between the sexes. Variability in sex differences in mortality in different times and places suggests cultural factors that may influence sex differences in mortality.

Cross-Cultural Comparisons

Infant mortality, that is, mortality during the first year of life, is higher for males than for females in almost all available data for different countries and different historical periods (Stolnitz 1956; United Nations 1978). In recent data the sex ratio for infant mortality in most countries fell in the range from 1.1 to 1.4, with the higher values being more common in countries with lower levels of infant mortality (data from United Nations 1978).

More detailed data suggest that different factors may influence sex differences in the first few months of life and in the later part of the first year. In general, sex mortality ratios are higher for the first few months of life, during which period higher mortality for males appears to be nearly universal (U.S. NCHS 1979c; D'Souza and Chen 1980; Kelly 1975; Miller 1981). This consistent sex difference in mortality suggests that biological factors probably contribute to greater male vulnerability in the first few months of life. It has been argued that a biological disadvantage for males extends to the prenatal period (Teitelbaum 1971). Mortality in the later months of gestation appears to be higher for males than for females, although a male excess has not been universally observed (Teitelbaum 1971;

Naeye et al. 1971; U.S. NCHS 1979c; Degenhardt and Michaelis 1977). For earlier months of in utero development there have been repeated reports of higher mortality for males, but these findings have been derived from studies that appear to have been biased due to methodological problems (Tietze 1948; Degenhardt and Michaelis 1977). Methodologically more reliable studies indicate that in earlier months of gestation mortality is as high for females as for males (Degenhardt and Michaelis 1977; Lauritsen 1976; Tietze 1948; WHO 1966). Thus, the female advantage in survival during the first few months after birth apparently does not extend to a consistent female advantage during in utero development.

During the second half of the first year male mortality is higher than female mortality in many countries such as the United States (U.S. NCHS 1979c), but female mortality is higher than male mortality in some regions. Higher female mortality is particularly well documented for South Asia, including northern India (Kelly 1975; Wyon and Gordon 1971) and one area of Bangladesh (D'Souza and Chen 1980). The female disadvantage is sufficient that infant mortality (mortality for the whole first year) is higher for females than males in northern India (Kelly 1975; Miller 1981). Factors contributing to higher infant mortality for females in this region appear to include less adequate nutrition and less medical care for females (Kelly 1975; Miller 1981; Wyon and Gordon 1971).

For ages above 1 year old, there is wide variation in the sex differences in mortality and strong evidence that cultural factors play an important role. For ages between 1 and 40, females have had higher death rates than males in many countries (figure 6-4). Higher mortality for females has been common in countries that were not industrialized or that were in the early stages of industrialization (Stolnitz 1956; El-Badry 1969; Tabutin 1978). In contemporary data, higher mortality for females at ages 1–44 has been documented for northern India (Wyon and Gordon 1971; Miller 1981; U.S. Bureau of the Census 1978), Nepal (U.S. Bureau of the Census 1979) and an area in Bangladesh (D'Souza and Chen 1980).

In the regions with higher female mortality, girls and young women have had higher mortality for a wide range of causes (Tabutin 1978; Preston 1976). The only major category with consistently higher mortality for males than for females at these ages appears to be violence (that is, accidents, suicide, and homicide). Female mortality has frequently been higher than male mortality for infectious diseases, including tuberculosis and diarrheal disease, for both of which susceptibility is substantially increased by inadequate nutrition. Genetic differences in immune resistance to infectious disease tend to favor females (Waldron 1976; Purtilo and Sullivan 1979). These observations suggest that women and girls have had higher rates of these infectious diseases because they have received less adequate nutrition than men and boys. This hypothesis is supported by historical evidence that,

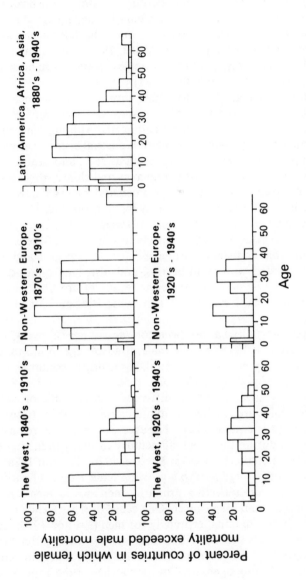

Source: Data from Stolnitz (1956). The West includes Western Europe and several other countries such as the United States and Australia.

Figure 6-4. Percent of Countries in Which Female Mortality Exceeded Male Mortality for Specified Ages

during industrialization in England, when many laboring-class families did not have adequate food, highly nutritious foods such as meat and cheese were more often given to the husband in order to maintain the health of the breadwinner, while the women had particularly inadequate diets (Oren 1973). Contemporary evidence indicates that in northern India and in Bangladesh, boys are better fed than girls (Miller 1981; Lindenbaum 1977). There is also evidence that in these regions males receive more medical care than females (Singh et al. 1962; Miller 1981).

Other factors that contribute to higher mortality for women are the physiological strains due to repeated pregnancy and lactation, as well as mortality directly due to pregnancy and childbirth. Although these factors have a minor impact on mortality in modern industrial societies, they are of considerable importance where resources such as food and health care are inadequate. Mortality associated with pregnancy and childbirth was responsible for about one-quarter of deaths for women under age 45 in two studies of rural regions in South Asia (Chen et al. 1974; Gordon et al. 1965). In one region this accounted for the entire excess of female deaths for ages 15–40 and in another region this accounted for about one-third of the excess. In summary, factors that contribute to higher mortality for girls and women include risks associated with child-bearing and also inadequate diet and health care in situations in which material resources are inadequate and males receive preferential treatment (Miller 1981; Preston 1976; El-Badry 1969).

Excess female mortality has not been common at ages above 45 (figure 6–4). However, even at these ages the male disadvantage in mortality has been much smaller in predominantly rural, agricultural countries than in industrial countries (figure 6–3) (Stolnitz 1956; Preston 1976; Waldron 1980a). One major reason is that sex differences in ischemic or coronary heart disease mortality are smaller in nonindustrial countries (figure 6–3) (Stamler et al. 1970; Ledermann 1964). The excess of male over female mortality has increased during the twentieth century in many industrial countries (Enterline 1961; Retherford 1975; Preston 1976; Tabutin 1978). This increase has been due in large part to a growing male excess for cardiovascular disease and cancer, reflecting primarily increases in men's rates of coronary heart disease and lung cancer. One cause of these trends has been an increase in cigarette smoking, especially for men (Preston 1970; Retherford 1975). Other factors that contribute to men's low rates of lung cancer and coronary heart disease in nonindustrial countries probably include low exposure to industrial carcinogens and low prevalence of the Type A behavior pattern (Bridbord et al. 1978; U.S. HEW Surgeon General 1979; Waldron 1980a).

Even among industrial countries, there has been considerable variation in the sex differences in mortality (figure 6–3) (Garros and Bouvier 1978; Preston 1970). Previous analyses indicate that differences in cigarette smoking and alcohol consumption contribute to the differences in sex mortality ratios among economically developed countries (Garros and Bouvier 1978; Preston 1970; Mulcahy et al. 1970; Ledermann 1964).

Trends in the United States

Analyses of historical data for the United States illustrate the importance of many of the trends and underlying causal factors discussed thus far. For the period before 1930 the data are incomplete, but available data indicate that from 1830 to 1930 the excess of female over male life expectancy fluctuated in the range from approximately 2 to 4 years, with occasional larger or smaller values (Haines 1977, 1979; U.S. NCHS 1980d). In 1930 the sex difference in life expectancy was 3.5 years (with a life expectancy of 58.1 years for males and 61.6 years for females) (U.S. NCHS 1980d). During the subsequent decades the sex difference in life expectancy increased considerably. By 1978, the sex difference in life expectancy was 7.7 years (with a life expectancy of 69.5 for males and 77.2 for females) (U.S. NCHS 1980d).

In the early part of the century sex differences in mortality were small at all ages (figure 6-5). Since 1930 sex mortality ratios have increased markedly, with most dramatic changes observed for teenagers and young adults. Enterline (1961) has analyzed the causes of the increase in sex differences in mortality for whites between 1930 and 1958. For the 15–24-year-old age group the most significant trends were a decrease in death rates for two causes for which young women had substantially higher mortality than young men in the early period, namely, tuberculosis and maternal mortality. For the 45–64-year-old age group, the major contributors to the increasing sex difference in mortality were a rise in men's death rates for heart disease and malignant neoplasms. As in other countries, the rise in men's death rates for coronary heart disease and lung cancer were of particular importance.

Since 1960, sex mortality ratios have continued to increase for most age groups, but sex mortality ratios appear to have stabilized for 45–64-year-olds (figure 6-5). An analysis by specific causes of death reveals complex underlying trends. The following brief description will focus on the contrast between the 15–24-year-old age group, for whom there has been a continuing substantial increase in sex mortality ratios, and the 45–54-year-old age group, for whom sex mortality ratios have been stable since 1960.

For the 15–24-year-old age group, the major cause of the increase in the sex mortality ratio since 1960 has been an increase in the proportion of deaths due to two causes with a large male excess: suicide and homicide. These two causes accounted for 10 percent of deaths in this age group in 1960 and 22 percent of deaths in 1976 (U.S. NCHS 1963, 1974c, 1978h). At the older ages, suicide and homicide are responsible for a much smaller

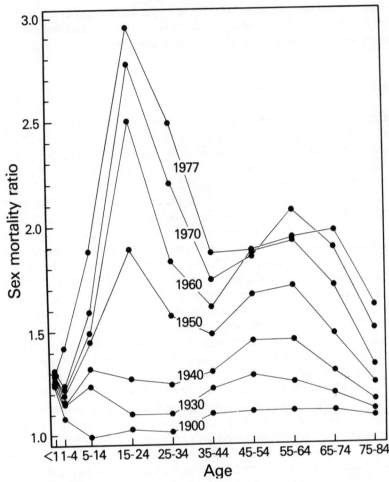

Sources: Data from Metropolitan Life (1974); U.S. NCHS (1978g).
Note: The sex mortality ratios, that is, the ratios of male-to-female death rates, increased
markedly from 1930 to 1977.

Figure 6–5. Sex Mortality Ratios for the United States, 1900–1970

fraction of total deaths, and the trends for suicide have been different, with
increases for women but decreases for men. These trends are reflected in the
decrease in sex mortality ratios for age-adjusted suicide rates beginning in
the late 1950s (figure 6–6).[5]

Another striking trend of importance at the older ages has been the
rapid decrease in sex mortality ratios for malignant neoplasms of the respiratory system due to a marked increase in women's death rates for this
cause (figure 6–6) (U.S. NCHS 1963, 1974c, 1976–1978). This recent

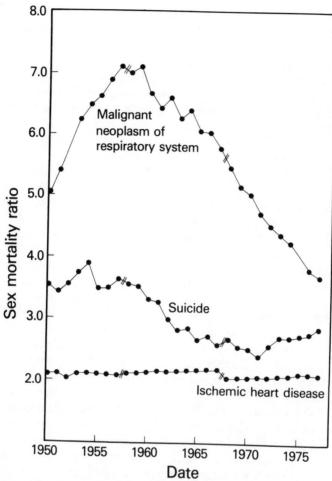

Sources: Data from U.S. NCHS (1974b, 1954–1977, 1976–1978); Havlik and Feinleib (1979).
Note: Hatchmarks indicate changes in the coding of specific causes of death.

Figure 6–6. Recent Trends in Sex Mortality Ratios for Malignant
Neoplasms of the Respiratory System, Suicide, and Ischemic
or Coronary Heart Disease in the United States

decrease in sex mortality ratios for malignant neoplasms of the respiratory
system follows a period of several decades in which the sex mortality ratios
rose due to increases in men's death rates (Burbank 1972; Gilliam et al.
1961). Burbank (1972) has presented evidence that the trends in sex differences in lung cancer correspond to the trends in sex differences in cigarette
smoking, with a latency of approximately thirty years between initial exposure to the carcinogens in cigarette smoke and the induced cancer deaths.

Another cause of death, of major importance at the older ages, is ischemic or coronary heart disease. Since 1950 the sex mortality ratio for coronary heart disease has been relatively stable (figure 6–6), in contrast to the increases during earlier decades (Moriyama et al. 1971; Havlik and Feinleib 1979; Enterline 1961). These trends have presumably been influenced by the trends in cigarette smoking. Since the mid–1930s there has been an increasing similarity in the prevalence of cigarette smoking among men and women, and this would be expected to result in decreasing sex mortality ratios for coronary heart disease over the last four decades (U.S. HEW, Surgeon General 1979; Burbank 1972). It seems clear that trends in cigarette smoking are not the only major factors influencing trends in sex differences in coronary heart disease since, instead of the predicted decline in sex mortality ratios over the last four decades, actual observed trends showed a stabilization beginning around 1950 (figure 6–6) (Moriyama et al. 1971).

Another factor of interest is the increase in women's labor force participation rates in the last five decades (U.S. DOL 1975). Employed women are more likely than housewives to display the hard-driving Type A behavior pattern, and this may reflect the effects of employment on women's behavior patterns (as well as the effects of behavior pattern on women's decisions to seek a job or to leave employment) (Haynes et al. 1978; Waldron 1978b). One study of coronary heart disease in women has shown an increased risk for clerical workers, although not for professional and managerial workers or blue-collar workers (Haynes and Feinleib 1980). The finding of increased risk for clerical workers was based on a relatively small number of cases and appears not to be confirmed by mortality statistics for several European countries (Anonymous 1980). These observations suggest the hypothesis that women's labor force participation may have had relatively little effect on their risk of coronary heart disease, at least thus far.[6] This hypothesis would be compatible with the observations that, despite the substantial increase in women's labor force participation rates in recent decades, women's coronary heart disease rates fell during the 1970s (Havlik and Feinleib 1979) and sex mortality ratios have been relatively stable since 1950 (figure 6–6).

The causes of recent trends in coronary heart disease are currently in considerable dispute (Havlik and Feinleib 1979), and it is therefore not surprising that the causes of the stabilization of the sex mortality ratio for coronary heart disease are not yet clear. As researchers are so fond of saying, more research is needed.

Conclusion

Cross-cultural and historical data provide evidence for both genetic and cultural influences on sex differences in mortality. For mortality due to coronary heart disease and mortality due to accidents and other violent causes, male death rates have consistently been at least as high as female death rates, but the magnitude of the sex differences has been highly variable. This suggests that biological differences between the sexes contribute to the greater vulnerability of males, but that cultural factors also have an important influence. Infant mortality has also been higher for males in almost all available data, and genetic factors can be assumed to contribute to this sex difference also.

Striking evidence that cultural factors have a substantial influence on sex differences in mortality is provided by two observations. First, in many nonindustrial or early industrial countries mortality has been higher for females than for males in childhood and at young adult ages. Second, in the United States and in many European countries the excess of male over female mortality has increased markedly during the twentieth century. Cultural factors that have contributed to these variations in sex differences include poor nutrition for females in many nonindustrial and early industrial countries, and sex differences in trends in cigarette smoking.

Conclusion

One conclusion that emerges clearly from this analysis of sex differences in morbidity and mortality is the importance of behavioral factors. Men's higher death rates are related to their higher rates of cigarette smoking, Type A behavior pattern, alcohol consumption, use of guns, and exposure to occupational hazards. Women's higher rates of doctor visits and more frequent reports of acute conditions probably reflect, in part, greater sensitivity of women to symptoms of illness and sex differences in typical patterns of response to stress.

Many of the behaviors that make an important contribution to the sex differences in mortality are strongly influenced by cultural characteristics, and this is reflected in the cross-cultural and historical variation in sex differences in mortality. The excess of male over female mortality has increased considerably in the twentieth century in industrial countries. This reflects trends such as increased cigarette smoking by men and probably also increased exposure to industrial carcinogens and increased prevalence of the Type A behavior pattern.

While this analysis points to the central role of behavioral differences between males and females and the importance of cultural influences on these behavioral differences, it is also true that genetically determined differences between the sexes contribute to sex differences in morbidity and mortality. For example, the more complex and demanding reproductive functions of women contribute to sex differences in morbidity. Female sex hormones may reduce women's risk of coronary heart disease. There may also be genetic influences on the sex differences in behavior that contribute to sex differences in health, although the extent of influence of genetic factors on sex differences in behavior is at present a highly controversial issue (see chapter 8). Finally, behavioral differences between the sexes may be indirectly influenced by sex differences in reproductive function. The fact that only women can bear and nurse children appears to have had a pervasive influence on the development of sex roles in many cultures. The pervasive differences in sex roles appear to have had the consequence that in industrial societies men have been exposed more than women to the hazards of occupational carcinogens, to pressures that lead to the development of the Type A behavior pattern, and to the risks associated with excessive alcohol consumption.

In the contemporary era, when men's and women's roles are becoming more flexible and when sex differences in behaviors such as cigarette smoking are diminishing, there is a possibility that sex differences in mortality may diminish in the future. On the pessimistic side, there is the possibility that changes in women's behavior may result in increasing mortality, as suggested, for example, by the increase in women's lung cancer rates since 1960. On the optimistic side, there is the possibility that decreases in men's mortality could result from social changes, such as a reduction in occupational hazards or a change in social conditions that might foster a reduction in Type A behavior.

Notes

1. Nathanson (1978) has reported similar results based on data from the U.S. Hospital Discharge Survey. It should be noted that this survey excludes hospitalizations in federal hospitals, including Veterans Administration and military hospitals, and thus underestimates male relative to female hospitalizations.

2. This estimate was derived as follows. The National Safety Council estimates that there were approximately 9,000 fatal work accidents that were non-motor-vehicle-accidents (National Safety Council 1980). Studies of national samples have found that men account for 77 percent of reported work accidents (U.S. NCHS 1976c, 1978c, 1979c). If men accounted for 85 percent of fatal work accidents, then men would have experienced about 6,300 more fatal work accidents than women. This is equivalent to an excess death rate of 5.9/100,000 males or one-third of the sex difference in death rates for other accidents (U.S. NCHS 1980a).

3. Gove and Hughes (1979) also showed that for the married adults in their sample, sex differences for the health measures were diminished and

not statistically significant when controls for socioeconomic status and the measures of nurturant role demands and psychological symptoms were introduced. This finding should not be interpreted as evidence that nurturant role demands, psychological distress, and socioeconomic status are entirely responsible for sex differences in reported illness. The absence of significant sex differences could be a statistical artifact due to controlling for factors that may elevate women's rates of illness while not controlling for factors that may elevate men's rates. Also, it is possible that the adjustment for sex differences in the index of nurturant role demands unintentionally included an adjustment for sex differences in the perception of illness, so the interpretation of analyses using the index of nurturant role demands must be made with caution. Specifically, if women more readily consider themselves to be "really sick," then this could contribute to their more frequent perception that when they are "really sick" they need to carry on with chores or are unable to rest or have no one to take care of them, and these items are primarily responsible for the sex difference in index of nurturant role demands. Sex differences in perception of illness could contribute to sex differences in another item in the index of nurturant role demands, namely, that the respondent often catches a "bug" from others in the household. Thus, sex differences in perception of illness may have contributed both to sex differences in the index of nurturant role demands and to the sex differences in the self-reported morbidity measures. For further discussion, see Verbrugge (1980) and Gove and Hughes (1980).

4. Data from other studies also indicate that, in general, women who have been employed have not had high mortality comparable to that observed for men (Waldron 1980b and unpublished data from Alameda County; Madigan 1957). (Madigan has argued that the finding that Roman Catholic Sisters had lower mortality than Roman Catholic Brothers, despite their similar occupations, was evidence of the contribution of genetic factors to sex differences in mortality. However, the Brothers smoked and drank more than the Sisters, and probably were socialized differently as children, and each of these differences probably contributed to their higher mortality.) Additional evidence that the employment of women is probably not a key determinant of sex differences in mortality is provided by the finding that countries with higher labor force participation rates for women have not had smaller sex differences in mortality (data from Segi et al. 1966; International Labor Office 1966).

5. The sex mortality ratio for age-adjusted suicide rates decreased until about 1970, but increased subsequently due to increasing suicide rates for young men.

The reader is cautioned that data concerning recent mortality trends presented in Johnson (1977) are in error. The rates Johnson presented for arteriosclerotic heart disease are roughly twenty times lower than the actual rates published by the National Center for Health Statistics (Havlik and Feinleib 1979) and appear to be rates for arteriosclerosis, a much more minor cause of death. The data given for suicide are also wrong (Waldron 1977).

6. It is possible that in the future women's coronary heart disease rates may show more evidence of the effects of the increase in women's employment (Waldron 1980b). A lag between behavioral change and change in coronary heart disease rates is to be expected because atherosclerosis is a slow process that becomes manifest as coronary heart disease only after a period of many years. In addition, it is possible that gradual social changes may result in increasing development of the Type A behavior pattern in girls and young women, with increasing risk of coronary heart disease for these women when they reach middle age and older ages.

References

Anonymous. 1980. Women, work and coronary heart disease. *Lancet* 2: 76–77.

Armitage, K.H., L.J. Schneiderman, and R.A. Bass. 1979. Response of physicians to medical complaints in men and women. *J.A.M.A.* 241: 2186–2187.

Bacon, M.K. 1973. Cross-cultural studies of drinking, pp. 171–194 in P.G. Bourne (ed.), *Alcoholism.* Academic Press, New York.

Berkman, P.L. 1975. Survival and a modicum of indulgence in the sick role. *Medical Care* 13:85–94.

Blum, R.H., assisted by L. Braunstein. 1967. Mind-altering drugs and dangerous behavior; alcohol. Appendix B, pp. 29–49, Task Force Report: Drunkenness. President's Commission on Law Enforcement and Administration of Justice, Government Printing Office, Washington, D.C.

Bock, E.W., and I.L. Webber. 1972. Suicide among the elderly—Isolating widowhood and mitigating alternatives. *J. Marriage Family* 34:24–31.

Bridbord, K., P. Decoufle, J.F. Fraumeni, et al. 1978. Estimates of the fraction of cancer in the U.S. related to occupational factors. Mimeo from National Cancer Institute, National Institute for Environmental Health Sciences, National Institute for Occupational Safety and Health, Washington, D.C.

Bunker, J.P. 1970. Surgical manpower. *New Engl. J. Med.* 282:135–144.

Burbank, F. 1972. U.S. lung cancer death rates begin to rise proportionately more rapidly for females than for males—A dose-response effect? *J. Chron. Dis.* 25:473–479.

Cahalan, D. 1970. Problem Drinkers. Jossey-Bass, San Francisco.

Chen, L., M.C. Gesche, S. Ahmed, et al. 1974. Maternal mortality in rural Bangladesh. *Studies in Family Planning* 5:334–341.

Chien, A., and L.J. Schneiderman. 1975. A comparison of health care utilization by husbands and wives. *J. Commun. Health* 1:118–126.

Cloninger, C.R., K.O. Christiansen, T. Reich, et al. 1978. Implications of sex differences in the prevalences of antisocial personality, alcoholism, and criminality for familial transmission. *Arch. Gen. Psychiatry* 35: 941–951.

Cohen, J.B. 1974. Sociocultural Change and Behavior Patterns in Disease Etiology: An Epidemiologic Study of Coronary Disease Among Japanese-Americans. Ph.D. thesis in Epidemiology, University of California, Berkeley.

Cruz-Coke, R. 1971. Genetic aspects of alcoholism. pp. 335–363 in Y. Israel and J. Mardones (eds.), *Biological Basis of Alcoholism.* Wiley-Interscience, New York.

Cutler, S.J., J. Scotto, S.S. Devesa, et al. 1974. Third National Cancer Survey—An overview of available information. *J. National Cancer Inst.* 53:1565–1575.

Decoufle, P. 1977. Statistically speaking. *J. Occup. Med.* 19:582–583.

Degenhardt, A., and H. Michaelis. 1977. Primäres Geschlects-verhältnis 125 ♂ zu 100 ♀?-Analyse eines Artefakts. *Zeitschrift fur Bevölkerungswissenschaft* 4:3–22.

Dentan, R.K. 1968. *The Semai—A Nonviolent People of Malaya.* Holt, Rinehart and Winston, New York.

D'Souza, S., and L.C. Chen. 1980. Sex differentials in mortality in rural Bangladesh. *Pop. Develop. Rev.* 6:257–270.

El-Badry, M.S. 1969. Higher female than male mortality in some countries of South Asia: A digest. *J. Am. Stat. Assoc.* 64:1234–1244.

Enterline, P.E. 1961. Causes of death responsible for recent increases in sex mortality differentials in the United States. *Milbank Mem. Fund Quart.* 39:312–338.

Epstein, F.H., T. Francis, N.S. Hayner, et al. 1965. Prevalence of chronic diseases and distribution of selected physiologic variables in a total community, Tecumseh, Michigan. *Am. J. Epidemiol.* 81:307–322.

Ferrence, R.G. 1980. Sex differences in the prevalence of problem drinking. *Res. Adv. Alcohol Drug Probl.* 5:69–124.

Fine, E.W., and P. Scoles. 1974. Alcohol, alcoholism and highway safety. *Publ. Health Rev.* 3:423–436.

Frank, K.A., S.S. Heller, D.S. Kornfeld, et al. 1978. Type A behavior pattern and coronary angiographic findings. *J.A.M.A.* 240:761–763.

Galambos, J.T. 1979. Cirrhosis. *Major Problems in Internal Medicine* 17: 1–376.

Garros, B. and M.H. Bouvier. 1978. Exces de la surmortalite masculine en France et causes medicales de deces. *Population* 6:1095–114.

Gilliam, A.G., B. Milmore, and J.W. Lloyd. 1961. Trends of mortality attributed to carcinoma of the lung. *Cancer* 14:622–628.

Gordon, J.E., S. Singh, and J.B. Wyon. 1976. Causes of death at different ages, by sex, and by season, in a rural population of the Punjab, 1957–1959. *Ind. J. Med. Res.* 53:906–917.

Gordon, T., W.B. Kannel, M.C. Hjortland, et al. 1978. Menopause and coronary heart disease—The Framingham Study. *Ann. Intern. Med.* 89:157–161.

Gove, W.R. 1973. Sex, marital status, and mortality. *Am. J. Sociol.* 79: 45–67.

Gove, W.R., and M. Hughes. 1979. Possible causes of the apparent sex differences in physical health: An empirical investigation. *Am. Sociol. Rev.* 44:126–146.

———. 1980. Sex differences in physical health and how medical sociologists view illness. *Am. Sociol. Rev.* 45:514–522.

Hackett, T.P., and N.H. Cassem. 1969. Factors contributing to delay in responding to the signs and symptoms of acute myocardial infarction. *Am. J. Cardiol.* 24:651–665.

Hackett, T.P., N.H. Cassem, and J.W. Raker. 1973. Patient delay in cancer. *New Engl. J. Med.* 289:14–20.

Haines, M.R. 1977. Mortality in nineteenth century America: Estimates from New York and Pennsylvania census data, 1865 and 1900. *Demography* 14:311–331.

———. 1979. The use of model life tables to estimate mortality for the United States in the late nineteenth century. *Demography* 16:289–312.

Hammond, E.C. 1966. Smoking in relation to the death rates of one million men and women. *Nat. Cancer. Inst. Monogr.* 19:127–204.

Hammond, E.C., and H. Seidman. 1980. Smoking and cancer in the United States. *Prev. Med.* 9:169–173.

Hammond, E.C., I.J. Selikoff, and H. Seidman. 1979. Asbestos exposure, cigarette smoking and death rates. *Ann. N.Y. Acad. Sci.* 330:473–490.

Harrison, J. 1978. Warning: The male sex role may be dangerous to your health. *J. Social Issues* 34:65–86.

Havlik, R.J., and M. Feinleib, eds. 1979. *Proceedings of the Conference on the Decline in Coronary Heart Disease Mortality.* U.S. HEW, NIH, Bethesda, Md.

Haynes, S.G., and M. Feinleib. 1980. Women, work and coronary heart disease: Prospective findings from the Framingham Heart Study. *Am. J. Publ. Health* 70:133–141.

Haynes, S.G., M. Feinleib, and W.B. Kannel. 1980. The relationship of psychosocial factors to coronary heart disease in the Framingham study: III. Eight-year incidence on coronary heart disease. *Am. J. Epidemiol.* 111:37–58.

Haynes, S.G., S. Levine, N. Scotch, et al. 1978. The relationship of psychosocial factors to coronary heart disease in the Framingham Study: II. Prevalence of coronary heart disease. *Am. J. Epidemiol.* 107: 384–402.

Hollnagel, H., and F. Kamper-Jørgensen. 1980. Utilization of health services by 40-year-old men and women in the Glostrup area, Denmark. *Danish Med. Bull.* 27:130–139.

International Labor Office. 1966. *Yearbook of Labor Statistics, 1965.* vol. 25. I.L.O., Geneva.

Jarvis, G.K., R.G. Ferrence, F.G. Johnson, et al. 1976. Sex and age patterns in self-injury. *J. Health Soc. Behav.* 17:145–154.

Jenkins, C.D. 1976. Recent evidence supporting psychologic and social risk factors for coronary disease. *New Engl. J. Med.* 294:1033–1038.

Johnson, A. 1977. Recent trends in sex mortality differentials in the U.S. *J. Human Stress* 3(1):22–32.

Kannel, W.B., and M. Feinleib. 1972. Natural history of angina pectoris in the Framingham Study. *Am. J. Cardiol.* 29:154–163.

Kannel, W.B., P. Sorlie, and P.M. McNamara. 1979. Prognosis after initial myocardial infarction: The Framingham Study. *Am. J. Cardiol.* 44: 53–59.

Kelly, N.U. 1975. Some Socio-Cultural Correlates of Indian Sex Ratios: Case Studies of Punjab and Kerala. Ph.D. thesis in South Asia Regional Studies, University of Pennsylvania, Philadelphia.

Kramer, M., E.S. Pollack, R.W. Redick, et al. 1972. *Mental Disorder/ Suicide.* Harvard University Press, Cambridge, Mass.

Lauritsen, J.G. 1976. Aetiology of spontaneous abortion. *Acta Obstet. Gynecol. Scand.* Suppl. 52:1–29.

Lawrence, J.J., and M.A. Maxwell. 1962. Drinking and socioeconomic status. Pp. 141–145 in D. Pittman and C. Snyder (eds.), *Society, Culture and Drinking Patterns.* Wiley, New York.

Ledermann, S. 1964. *Alcool, Alcoolisme, Alcoolisation.* vol. 2, Travaux et Documents, Cahier 41. Presses Universitaires de France.

Lelbach, W.K. 1974. Organic pathology related to volume and pattern of alcohol use. *Research Adv. in Alcohol and Drug Programs* 1:93–198.

Leo, R.G., A.H. Dysterheft, and G.G. Merkel. 1957. Health profile of lumber mill employees. *Ind. Med. Surg.* 26:377–379.

Lewis, C.E., M.A. Lewis, A. Lorimer, et al. 1977. Child-initiated care: The use of school nursing services by children in an "adult-free" system. *Pediatrics* 60:499–507.

Lindenbaum, S. 1977. The "last course": Nutrition and anthropology in Asia. Pp. 141–155 in T.K. Fitzgerald (ed.), *Nutrition and Anthropology in Action.* Van Gorcum Press, Amsterdam.

Maccoby, E.E., and C.N. Jacklin. 1974. *The Psychology of Sex Differences.* Stanford University Press, Stanford.

MacMahon, B., P. Cole, and J. Brown. 1973. Etiology of human breast cancer: A review. *J. Natl. Cancer Inst.* 50:21–42.

MacMahon, B., S. Johnson, and T.F. Pugh. 1963. Relation of suicide rates to social conditions. *Publ. Health Rep.* 78:285–293.

Maddox, G.L. 1964. Self-assessment of health status. *J. Chron. Dis.* 17: 449–460.

Madigan, F.C. 1957. Are sex mortality differentials biologically caused? *Milbank Mem. Fund Quart.* 35:202–223.

Marston, M.V. 1970. Compliance with medical regimens: A review of the literature. *Nursing Res.* 19:312–322.

Matthews, K.A., and D.S. Krantz. 1976. Resemblances of twins and their parents in pattern A behavior. *Psychosomat. Med.* 38:140–144.

McCranie, E.W., A.J. Horowitz, and R.M. Martin. 1978. Alleged sex-role stereotyping in the assessment of women's physical complaints: A study of general practitioners. *Soc. Sci. Med.* 12:111–116.

McGill, H.C. and M.P. Stern. 1979. Sex and atherosclerosis. *Atherosclerosis Rev.* 4:157–242.

Mechanic, D. 1976. Sex, illness, illness behavior and the use of health service. *J. Human Stress* 2(4):29–40.

Mellinger, G.D., M.B. Balter, D.I. Manheimer, et al. 1978. Psychic distress life crisis, and use of psychotherapeutic medications. *Arch. Gen. Psychiatry* 35:1045–1052.

Menck, H.R., and B.E. Henderson. 1976. Occupational differences in rates of lung cancer. *J. Occup. Med.* 18:797–801.

Metropolitan Life Insurance Company. 1974. Sex differentials in mortality. *Statistical Bull.* 55:2–5.

Miller, B.D. 1981. *The Endangered Sex: Neglect of Female Children in Rural North India.* Cornell University Press, Ithaca, N.Y.

Moriyama, I., D. Krueger, and J. Stamler. 1971. *Cardiovascular Diseases in the United States.* Harvard University Press, Cambridge, Mass.

Moss, A.J., B. Wynar, and S. Goldstein. 1969. Delay in hospitalization during the acute coronary period. *Am. J. Cardiol.* 24:659.

Mulcahy, R., J.W. McGilvray, and N. Hickey. 1970. Cigarette smoking related to geographic variations in coronary heart disease mortality and to expectation of life in the two sexes. *Am. J. Publ. Health* 60: 1515–1521.

Mushinski, M.H., and S.D. Stellman. 1978. Impact of new smoking trends on women's occupational health. *Prev. Med.* 7:349–365.

Naeye R.L., L.S. Burt, D.L. Wright, et al. 1971. Neonatal mortality, the male disadvantage. *Pediatrics* 48:902–906.

Nathanson, C.A. 1977*a*. Sex, illness and medical care—A review of data theory and method. *Soc. Sci. Med.* 11:13–25.

———. 1977*b*. Sex roles as variables in preventive health behavior. *J. Commun. Health* 3:142–155.

———. 1978. Sex roles as variables in the interpretation of morbidity data: A methodological critique. *Int. J. Epidemiol.* 7:253–262.

National Safety Council. 1980. *Accident Facts, 1980.* National Safety Council, Chicago.

NIOSH. 1977. National Institute for Occupational Safety and Health. *National Occupational Hazard Survey,* vol. 3. *Survey Analysis and Supplemental Tables.* U.S. HEW, NIOSH, Cincinnati, Ohio.

Nordøy, A., A. Aakvaag, and D. Thelle. 1979. Sex hormones and high density lipoproteins in healthy males. *Atherosclerosis* 34:431–436.

Oren, L. 1973. The welfare of women in laboring families, England, 1860–1950. *Feminist Studies* 1:107–126.

Otto, R. 1979. Negative and positive life experience among men and women in selected occupations, symptom awareness and visits to the doctor. *Soc. Sci. Med.* 131:151–164.

Peck, M.L., and A. Schrut. 1971. Suicidal behavior among college students. *HSMHA Health Reports* 86:149–156.

Preston, S.H. 1970. An international comparison of excessive adult mortality. *Pop. Studies* 24:5–20.

———. 1976. *Mortality in National Populations.* Academic Press, New York.

Purtillo, D.T., and J.L. Sullivan. 1979. Immunological bases for superior survival of females. *Am. J. Dis. Child.* 133:1251–1253.

Retherford, R.D. 1975. *The Changing Sex Differential in Mortality.* Studies in Population and Urban Demography no. 1. Greenwood Press, Westport, Conn.

Robinson, J.P., P.E. Converse, and A. Szalai. 1972. Everyday life in twelve countries. Pp. 113–114 in A. Szalai (ed.), *The Use of Time* Mouton, The Hague.

Rosenman, R.H., N.O. Borhani, R. Rahe, et al. 1976. Heritability of Personality and Behavior Pattern. *Act. Genet. Med. Gemellol.* (Roma) 25:221–224.

Schneidman, E.S., and N.L. Farberow. 1961. Statistical comparisons between attempted and committed suicides. Pp. 19–47 in N.L. Farberow and E.S. Schneidman (eds.), *The Cry for Help.* McGraw-Hill, New York.

Segi, M., M. Kurihara, and Y. Tsukahara. 1966. *Mortality for Selected Causes in 30 Countries 1950–1961.* Tokyo: Kosei Tokei Kyokai.

Selikoff, I.J., E.C. Hammond, and H. Seidman. 1979. Mortality experience of insulation workers in the United States and Canada, 1943–1976. *Ann. N.Y. Acad. Sci.* 30:91–115.

Selikoff, I.J., and D.H.K. Lee. 1978. *Asbestos and Disease.* Academic Press, New York.

Selikoff, I.J., H. Seidman, and E.C. Hammond. 1980. Mortality effects of cigarette smoking among amosite asbestos factory workers. *J. Nat. Cancer Inst.* 65:507–513.

Shurtleff, D. 1970. Some characteristics related to the incidence of cardiovascular disease and death, Framingham Study, 16-year follow-up. In W.B. Kannel and T. Gordon (eds.), *The Framingham Study, An Epidemiological Investigation of Cardiovascular Disease,* Sept. 26, Dept. HEW, Washington D.C., U.S. GPO.

Silverstein, B., S. Feld, and L.T. Kozlowski. 1980. The availability of low nicotine cigarettes as a cause of cigarette smoking among teenage females. *J. Health Soc. Behav.* 21:383–388.

Simon, A.B., M. Feinleib, and H.K. Thompson, Jr. 1972. Components of delay in the pre-hospital setting of acute myocardial infarction. *Am. J. Cardiol.* 30:475.

Singh, S., J.E. Gordon, and J.B. Wyon. 1962. Medical care in fatal illnesses of rural Punjab population. *Ind. J. Med. Res.* 50:865–960.

Stamler, J., R. Stamler, and R.B. Shekelle. 1970. Regional differences in prevalence, incidence and mortality from atherosclerotic coronary heart disease. In J.H. deHass, H.C. Hemker and H.A. Sneller (eds.), *Ischaemic Heart Disease.* Williams and Williams, Baltimore.

Stengel, E., and N.G. Cook. 1958. *Attempted Suicide.* (Maudsley Monographs, no. 4. Chapman and Hall, London.

Stolnitz, G.J. 1956. A century of international mortality trends: II. *Pop. Studies* 10:17–42.

Strong, J.P., C. Restrepo, and M. Guzman. 1978. Coronary and aortic atherosclerosis in New Orleans: II Comparison of lesions by age, sex and race. *Lab. Investig.* 39:364–369.

Tabutin, D. 1978. La surmortalité féminine en Europe avant 1940. *Population,* no. 1:121–148.

Taylor, D.G., L.A. Aday, and R. Andersen. 1975. A social indicator of access to medical care. *J. Health Soc. Behav.* 16:39–49.

Teitelbaum, M.S. 1971. Male and female components of perinatal mortality: International trends, 1901–63. *Demography* 8:541–548.

Tessler, R, D. Mechanic, and M. Dimond. 1976. The effect of psychological distress on physician utilization: A prospective study. *J. Health Soc. Behav.* 7:353–364.

Tietze, C. 1948. A note on the sex ratio of abortions. *Human Biol.* 20:156–160.

Uhlenhuth, E.H., R.S. Lipman, M.B. Balter, et al. 1974. Symptom intensity and life stress in the city. *Arch. Gen. Psychiatry* 31:759–764.

United Nations. 1978. *U.N. Demographic Yearbook.* New York: United Nations.

U.S. Bureau of the Census. 1975. *Statistical Abstract of the U.S.* (96th ed.) Washington, D.C.: U.S. GPO.

———. 1977. *Social Indicators, 1976.* U.S. GPO, Washington, D.C.

———. 1978. *Country Demographic Profiles: India.* U.S. GPO, Washington, D.C.

———. 1979. *Country Demographic Profiles: Nepal.* U.S. GPO, Washington, D.C.

———. 1980. *Country Demographic Profiles: Pakistan.* U.S. GPO, Washington, D.C.

U.S. DOL. 1975. Department of Labor. *1975 Handbook on Women Workers.* Women's Bureau Bulletin no. 297. U.S. GPO, Washington, D.C.

———. 1980. *An Interim Report to Congress on Occupational Diseases.* U.S. GPO, Washington, D.C.

U.S. FBI. 1964, 1972, 1974, and 1980. Federal Bureau of Investigation. Crime in the U.S. 1963, 1971, 1973, and 1979. Uniform Crime Reports for the U.S. U.S. Department of Justice, Washington, D.C.

U.S. HEW, NCI. 1977. National Cancer Institute. *Cigarette Smoking among Teenagers and Young Women.* DHEW publ. no. (NIH) 77-1203, U.S. GPO, Washington, D.C.

U.S. HEW, Surgeon General. 1979. Office on Smoking and Health. *Smoking and Health—A Report of the Surgeon General.* DHEW publ. no. (PHS) 79-50066. U.S. GPO, Washington, D.C.

U.S. NCHS. 1954–1977. HEW, National Center for Health Statistics. *Vital Statistics of the U.S., 1950–1973.* U.S. GPO, Washington, D.C.

———. 1963. *Vital Statistics of the U.S., 1960,* vol. II. *Mortality.* U.S. GPO, Washington, D.C.

———. 1965. Coronary heart disease in adults, U.S., 1960–2. *Vital and Health Stat.* series 11, no. 10.

———. 1967. Three views of hypertension and heart disease. *Vital and Health Stat.* series 2, no. 22.

———. 1973a. Net differences in interview data on chronic conditions and information derived from medical records. *Vital and Health Stat.* series 2, no. 57.

———. 1973b. Quality control and measurement of nonsampling error in the Health Interview Survey. *Vital and Health Stat.* series 2, no. 54.

———. 1974a. Current Estimates from the Health Interview Survey U.S., 1973. *Vital and Health Stat.* series 10, no. 95.

———. 1974b. Mortality trends for leading causes of death, U.S., 1950–1969. *Vital and Health Stat.* series 20, no. 16.

———. 1974c. *Vital Statistics of the U.S., 1970,* vol. II. *Mortality.* U.S. GPO, Washington, D.C.

———. 1974d. Hypertension and hypertensive heart disease in adults, U.S., 1960–2. *Vital and Health Stat.* series 11, no. 13.

———. 1976–1978. Advance report—Final mortality statistics 1974–1977.

Monthly Vital Stat. Report 24–11, 25–11 and 26–12, supplements.

U.S. NCHS. 1976*a*. Hospital discharges and length of stay: Short-stay hospitals, U.S., 1972. *Vital and Health Stat.* series 10, no. 107.

———. 1976*b*. Blood pressure of persons 6–74 years of age in the United States. *Advancedata* no. 1.

———. 1976*c*. Persons injured and disability days by detailed type and class of accident, U.S., 1971–2. *Vital and Health Stat.* series 10, no. 105.

———. 1976*d*. Hypertension: U.S., 1974. *Advancedata* no. 2.

———. 1976*e*. Differentials in health characteristics by marital status, U.S., 1971–2. *Vital and Health Stat.* series 10, no. 104.

———. 1977*a*. Current estimates from the Health Interview Survey, U.S., 1975. *Vital and Health Stat.* series 10, no. 115.

———. 1977*b*. Current estimates from the Health Interview Survey, U.S., 1976. *Vital and Health Stat.* series 10, no. 119.

———. 1977*c*. A comparison of levels of serum cholesterol of adults 18–74 years of age in the U.S. in 1960–1962 and 1971–1974. *Advancedata* no. 5.

———. 1977*d*. Limitation of activity due to chronic conditions. U.S., 1974. *Vital and Health Stat.* series 10, no. 11.

———. 1978*a*. Current Estimates from the Health Interview Survey U.S., 1977. *Vital and Health Stat.* series 10, no. 126.

———. 1978*b*. The National Ambulatory Medical Care Survey: 1975 summary. *Vital and Health Stat.* series 13, no. 33.

———. 1978*c*. Episodes of person injured, U.S., 1975. *Advancedata* no. 18.

———. 1978*d*. Blood pressure levels of persons 6–74 years, U.S., 1971–4. *Vital and Health Stat.* series 11, no. 203.

———. 1978*e*. Access to ambulatory health care: United States 1974. *Advancedata* no. 17.

———. 1978*f*. Disability days, U.S. 1975. *Vital and Health Stat.* series 10, no. 118.

———. 1978*g*. Provisional Statistics—Annual summary for the U.S., 1977. *Monthly Vital Stat. Report* 26, no. 13.

———. 1978*h*. Facts of Life and Death. DHEW publ. no. (PHS) 79–1222. U.S. GPO, Washington, D.C.

———. 1979*a*. Advance report—Final mortality statistics, 1977. *Monthly Vital Stat. Report* 28, no. 1, suppl.

———. 1979*b*. Current estimates from the Health Interview Survey, U.S., 1978. *Vital and Health Stat.* series 10, no. 130.

———. 1979*c*. *Vital Statistics of the U.S., 1975,* vol. II. *Mortality.* U.S. GPO, Washington, D.C.

———. 1979*d*. Acute conditions, U.S., 1977–8. *Vital and Health Stat.* series 10, no. 132.

U.S. NCHS. 1980*a*. Advance report: Final mortality statistics, 1978. *Monthly Vital Stat. Report* 29, no. 6, suppl. 2.

———. 1980*b*. The national ambulatory medical care survey, U.S., 1977. *Vital and Health Stat.* series 13, no. 44.

———. 1980*c*. Basic data on health care needs of adults ages 25–74 years, 1971–5. *Vital and Health Stat.* series 11, no. 218.

———. 1980*d*. Life Tables. *Vital Statistics of the U.S., 1978*, vol. 2, sect. 5. U.S. GPO, Washington, D.C.

Verbrugge, L.M. 1976*a*. Sex differentials in morbidity and mortality in the United States. *Social Biol.* 23:275–296.

———. 1976*b*. Females and illness: Recent trends in sex differences in the United States. *J. Health Soc. Behav.* 17:387–403.

———. 1978. Complaints and diagnoses: Sex differences in the vocabulary and attribution of illness. Paper presented at the American Public Health Association meeting, Los Angeles, October 1978.

———. 1980. Comment of Gove and Hughes, "Possible causes of the apparent sex differences in physical health." *Am. Sociol. Rev.* 45:507–513.

Waldron, I. 1976. Why do women live longer than men? *Soc. Sci. Med.* 10:349–362.

———. 1977. Sex mortality differentials. *J. Human Stress* 3(2):46.

———. 1978*a*. Sex differences in the coronary-prone behavior pattern. Pp. 199–205 in T.M. Dembroski, S.M. Weiss, J.L. Shields, et al. (eds.), *Coronary-prone Behavior*. Springer-Verlag, New York.

———. 1978*b*. The Coronary-prone behavior pattern, blood pressure, employment and socioeconomic status in women. *J. Psychosom. Res.* 22:79–87.

———. 1978*c*. Coronary heart disease and male sex hormones. *Brit. Med. J.* 1:1485.

———. 1980*a*. Sex differences in longevity. Pp. 163–182 in S.G. Haynes and M. Feinleib (eds.), *Second Conference on the Epidemiology of Aging*. U.S. DHHS, NIH publ. no. 80–969. U.S. GPO, Washington D.C.

———. 1980*b*. Employment and women's health: An analysis of causal relationships. *Internat. J. Health. Serv.* 10:435–454.

———. 1980*c*. Mortality and sex difference. *AM. J. Dis. Child.* 134:999.

Waldron, I., J. Herold, D. Dunn, et al. 1981. Reciprocal effects of health and labor force participation in women—Evidence from two longitudinal studies. Paper presented at the Fourteenth Annual Meeting of the Society for Epidemiological Research, Snowbird, Utah, June 17–19, 1981.

Waldron, I., A. Hickey, C. McPherson, et al. 1980. Type A Behavior Pattern: Relationships to variation in blood pressure, parental characteristics, and academic and social activities of students. *J. Human Stress* 6(1):16–27.

Weinblatt, E., S. Shapiro, and C.W. Frank. 1973. Prognosis of women with newly diagnosed coronary heart disease: A comparison with course of disease among men. *Am. J. Public Health* 63:577–593.

Williams, R.R., and J.W. Horn. 1977. Association of cancer sites with tobacco and alcohol consumption and socioeconomic status of patients: Interview study from the Third National Cancer Survey. *J. Natl. Cancer Inst.* 58:525–547.

Winokur, G. 1967. X-borne recessive genes in alcoholism. *Lancet* 2:466.

Wolfgang, M.E. 1958. *Patterns in Criminal Homicide.* University of Pennsylvania Press, Philadelphia.

Wolinsky, F. 1978. Assessing the effects of predisposing, enabling and illness-morbidity characteristics on health service utilization. *J. Health Soc. Behav.* 19:384–396.

Worden, J.W., and A.D. Weisman. 1975. Psychosocial components of lagtime in cancer diagnosis. *J. Psychosom. Res.* 19:69.

World Health Organization. 1966. Standardization of procedures for chromosomes studies in abortion. *Bulletin WHO* 34:765–782.

——. 1975. *World Health Statistics Annual, 1972.* vol. 1. *Vital Statistics and Causes of Death.* Geneva: World Health Organization.

Wyon, J.B., and J.E. Gordon. 1971. *The Khanna Study.* Harvard University Press, Cambridge, Mass.

7

Sexual Differentiation as a Model for Genetic and Environmental Interaction Affecting Physical and Psychological Development

C. Dawn Delozier
and *Eric Engel*

In keeping with the subject of sociobiology and the interactions of nature and nurture in the determination of human behavior, we have attempted to find a topic that could serve as a general model for such studies. Certainly, each branch of science, whether medical or humanistic, can learn much from the others, and we must begin to work more closely together if we are to unravel some of the mysteries of our species! We have decided to review some aspects of what may well be hormonally mediated behavior patterns, under some degree of prenatal control or imprinting effects. These patterns are those generally referred to as adult sexual dimorphic behavior. In our attempt to clarify some of the aspects of this topic, we will discuss it in three parts, using as examples several genetically determined disorders of sexual development in which prenatal developmental influences may have effects on later sexual dimorphic behavior characteristics. First, we present a brief outline of normal sexual development and the factors thought to control it; second, we will provide brief descriptions of the biological and phenotypical features of three selected genetic disorders; in the third section we will review some studies of the sexual dimorphic behavioral characteristics of individuals affected with such disorders. It is our belief that one can utilize these genetic "experiments of nature," in which we know the biological bases of the problems, to investigate some of the predispositions and determinants of human behavior.

Steps in Normal Prenatal Sexual Development

The first evidence of sexual development in the human fetus can be seen at three weeks of fetal life, in the formation of a "genital ridge." By six weeks of gestation a fetal gonad is present, though at this point the eventual ovary

117

or testis is completely "indifferent"—it is capable of differentiation in either sexual direction, depending on which genetic signals influence it. At this time the fetus also has two sets of internal sexual ducts: the Wolffian, which are the precursors of the male internal genitalia, and Müllerian, which are feminine in nature. Despite the presence of both types of tissues in the 6–7-week fetus, it seems that development in the female direction is the more generalized occurrence; in the absence of specific male-determining factors, both the internal and external sexual features continue feminine differentiation. In the case of the chromosomal female (46,XX karyotype), the anlages of internal genitalia become fallopian tubes, uterus, and upper portion of the vagina whereas the originally indifferent "genital tubercle" becomes clitoris, genital folds, and the labia minora. The first "genital swelling" becomes the labia majora (figure 7–1).

For male development to proceed, on the other hand, there must be a stimulation of the Wolffian system, which develops into the vas deferens, seminal vesicles, and epididymis. Externally, the genital tubercle becomes the glans of the penis, the folds unite as the penile shaft, and the genital swelling develops into a scrotum. All this is completed between about ten and fifteen weeks of fetal life, though these developments may certainly be modified later on by specific hormonal and other factors, both prenatally and postnatally [1–3].

So what are the specific determinants of fetal physical sexual development? The original step comes at fertilization, with the determination of the chromosomal sex of the individual as female, that is, karyotype 46,XX, or as male, 46,XY. The Y chromosome is one component essential for male sexual differentiation; its presence is so strong in this regard that even individuals with a chromosomal disorder where there are four X chromosomes and only one Y (49,XXXXY) are phenotypically differentiated in the masculine direction. Our knowledge of what factors are determined directly by the Y chromosome is as yet somewhat unclear, but it seems that both a specific testicular determining agent or activator (gene), and a surface antigen of the immune system, known as the HY antigen, are produced as a result of genes located on the Y chromosome. However, the Y chromosome itself is not sufficient for complete male development; one must have proper function not only of the sex-determining genes on the X and Y chromosomes but also of genes located on the autosomes (the other 22 pairs besides X and Y, which primarily determine body, or somatic, characteristics), such as those that aid in the metabolism of male hormones. Though the state of our knowledge of this subject is not complete, there appear to be three major factors that direct male sexual differentiation (see figure 7–2): (1) *the HY antigen,* evidently produced by a gene(s) on the Y chromosome; (2) male androgenic hormones, primarily *testosterone,* produced by the fetal testes; and (3) the *Müllerian inhibiting factor,* produced also by the fetal testes [2,4].

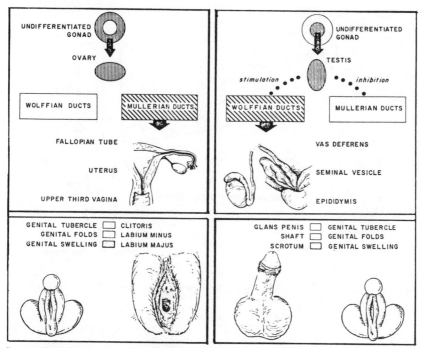

Source: Daniel Federman, *Abnormal Sexual Development,* Philadelphia: Saunders, 1967, pp. 2 and 3. Reprinted with permission.

Note that distinct internal duct primordia coexist in both sexes; that normal development involves growth of one system plus regression of the other; and that a testicular substance or substances play the major role in this process. In contrast, the external genitalia develop in a continuous transformation from an anlage common to embryos of both sexes.

Figure 7-1. Schematic outline of Normal (A) Female and (B) Male Differentiation

The first of these three substances to act is apparently the HY antigen, which at 6-7 weeks of fetal development is one of the stimulators for the emergence of a testis from the gonadal primordium that would otherwise have become an ovary. The HY antigen is a weak transplantation antigen of the immune system that can be detected in all male body cells with the exception of immature sperm [5]; the study of the HY antigen and its functions is one of the most active areas of sexual development research today.

Once the testis becomes functional, it produces and secretes two classes of hormones, beginning at about 8-9 weeks of fetal development, that act initially on structures adjacent to the testes, and later on "target" cells in other parts of the body, including the brain (figure 7-3): (1) *testosterone* (and other androgens), which when activated become directly responsible for masculinization of the external genitalia; and (2) *Müllerian inhibiting factor,* a large protein that causes regression of the primitive female ducts

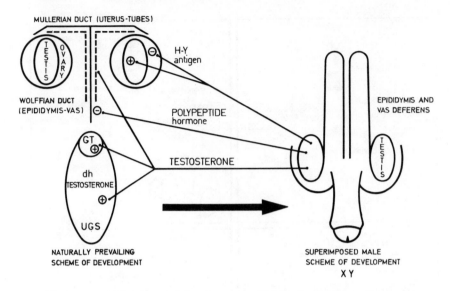

Figure 7-2. The Three Major Factors of Male Genital Development

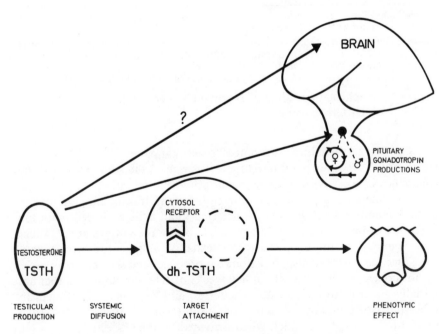

Figure 7-3. Brain Sexual Differentiation

that are present from the earliest stages of sexual development; by 10 weeks these structures are virtually absent [4]. Therefore, though steroid production and action continue throughout fetal development as well as later life, the main morphological differentiation of the sexual structures is determined during the first three months of pregnancy.

As might be imagined with such a complex system, there are numerous chances for error; the absence or nonfunction of any of several influences can lead to various degrees of intersexuality, with persisting feminine derivatives or incomplete masculinization of male structures. Indeed, we know of the existence of several of the intersex disorders that one might predict could result from errors in known biochemical pathways of steroid synthesis. Take for example the enzymatic secretion and function of testosterone, for the formation of which there are at least four intracellular events, each of which apparently requires proper function of several autosomally located genes and their products. If any one of these enzymes, cofactors, or activators is missing, there will be a deficit of testosterone, with resulting deficient masculinization of the individual. And the story of testosterone is even more complex than this, since to become effective on the genitalia, an enzymatic reduction of dihydrotestosterone is required. The next step is the binding of dihydrotestosterone to a cytoplasmic receptor protein at the end-organ site, such as the cells of the genital tubercle. This protein-hormone complex is then transported into the nucleus of the cell, where it may direct the formation of new enzymes and proteins needed specifically for male physical development [2].

Selected Genetic Disorders of Sexual Development

After this review of some of the major steps of male sexual development, we move to an explanation of the biological and phenotypic features of three genetic disorders that alter sexual morphology. Comparison of the physical and behavioral characteristics manifested by affected individuals may help us elucidate the specific functions and effects of the various male-determining factors of which we are aware.

Testicular feminization (or the androgen insensitivity syndrome) is an inherited disorder in which there is a lack of testosterone responsivity in all targ t cells, such as those of the external genitalia, probably because the cytoplasmic receptor at the cell level is abnormal [5]. Earlier we mentioned the strength of the Y chromosome in determination of masculine factors. Though this is without doubt true, the gene change that causes testicular feminization is apparently overriding. We recently studied a patient who had this syndrome in spite of the presence of an XYZ karyotype [6]. As a result, a chromosomal male, who has internal testes and masculine structures that produce normal amounts of testosterone and HY antigen, has the external genitalia of a female. Since the testosterone is never "seen" by the body cells (whether of the genitals or the brain), the developing urogenital system responds in all ways as if it were not present. Since the HY antigen and Müllerian inhibiting factors are present as in a normal male, the internal structures differentiate in masculine fashion. However, external genitalia are feminine, as will be secondary characteristics at puberty, since masculinization requires testosterone. As is the case with all normal males, some amount of female estrogen hormone is present, apparently enough to cause secondary sexual characteristics to develop. And since these women do not have the antagonizing effects of testosterone to modify the influences of estrogens, they may even be voluptuous in their physical development, a characteristic sometimes referred to as "florid feminization." Because of the absolute lack of response to testosterone, this is one of the sexual development defects which has been most useful in elucidating what specific effects prenatal androgens may have, in regard to both physical and psychosexual development.

Another disorder that involves androgenic action, actually a disorder of the adrenal system due to malproduction of the hormone cortisone, is congenital adrenal hyperplasia, commonly known as the adrenogenital syndrome (AGS). The end result of this inherited enzyme deficiency is an overproduction of androgens causing varying degrees of excessive masculinization of external and even internal genitalia. Though this disorder may affect either males or females, it is the genetic females who have genital ambiguity, which may range in severity from mild clitoral hypertrophy to masculinization marked enough that the individual is assigned a male sex at birth. For life as a female, cortisone replacement therapy to prevent buildup of androgenic by-products (and for other health reasons as well) is necessary. Corrective clitoral and vaginal surgery is often needed, but can be performed early and with excellent results. If the masculinized female is to live as a male, he will require not only cortisone and testosterone replacement, but also removal of female internal structures including ovaries. We will see later how this androgenization may have effects on behavioral patterns in adulthood. Keep in mind that in a female with the adrenogenital syndrome, only one of the three major elements of human male sexual development is present—there are androgens, but no other Y-determined factors such as HY antigen, nor is there Mullerian inhibiting factor present.

In our third example we find an excess of at least one and possibly two factors, but no incongruity between genetic and morphological sex. Affected individuals are genetic males in whom there has been an error during spermatogenesis so that there is an extra Y chromosome and a karyotype 47,XYY. There is therefore an excess of male factors such as the HY antigen and possibly testosterone [5]. The important point here is not at all one of sexual ambiguity, as the individuals are completely masculine in phenotype, but a question of what effects the extra Y-determined factors might have on sexual and other types of behavior patterns. This disorder has received more than its share of negative publicity, due to a regrettable early association with criminality. We now know that though the incidence of the XYY male is indeed higher in the mental-penal population than in the male population at birth [7, 8], this association is still rather uncommon. We will return to this question in a discussion of behavioral characteristics.

Before we turn to behavioral studies of these disorders, we should review the masculinizing parameters present in each of these genetic conditions (see table 7-1). First of all, present in normal male development are testes, HY antigen, androgens, and a small amount of estrogen. In the feminizing testes syndrome, in which affected individuals are phenotypically normal females though genetic males, there are also testes, HY antigen, and at least the normal male amount of estrogen. Androgens, on the other hand, are effectively absent. The XYY individual, a genetic and phenotypic male who demonstrates normal testes (generally) and male levels of estrogen, has levels of HY antigen and possibly of androgens that are elevated. These can be compared with the normal female, who has no testes, no HY antigen, only a small amount of testosterone, but who of course has the full feminine complement of estrogen. The female with adrenogenital syndrome, like the normal female, has no testes or HY antigen, but has normal female estrogen levels. In this case, the disorder results from an excess of androgens. With these factors in mind, we move to considerations of the behavioral characteristics of individuals with these disorders of sexual development.

In our evaluation of some of the biological aspects of the acquisition of gender role and identity, we will utilize the three disorders already outlined to illustrate the effects, proved or potential, that prenatal sex-determining factors may have on later sexual dimorphic behavior. Though there is no doubt that the influences by which the postnatal familial and societal environments shape sexual identity acquisition are numerous and complex, keep

Table 7-1

Some Parameters of Normal and Abnormal Sex Development

	Testes	H-Y	Androgens	Estrogens
Normal male	+	+	+	(+)
Feminizing testis	+	+	0	(+) to +
XYY	+	+ +	+ to + +	(+)
Normal female	0	0	(+)	+
♀ Adrenal hyperplasia	0	0	+ +	(+)

Note: ( ) = relatively weak value.

in mind the corollary already introduced: that postnatal societal factors act on an individual who already has certain sexual behavioral predispositions as a result of hormonal influences acting prenatally. Much evidence from animal studies supports this concept of prenatal "imprinting" of the developing brain. At the extreme end of the spectrum of hormonal influences, we know that complete sex reversal is possible in fish and amphibians when the sex-appropriate hormone is dissolved in water where the larvae exist [9]. The sex-reversed animals can even breed successfully!

On the more subtle subject of behavior in normal-sexed individuals, numerous studies on various mammalian and lower species have supported the concept of imprinting of some sexual dimorphic behaviors. For example, studies by Shapiro and Goldman [10], among others, have shown that appropriate sexual and maternal role playing in the adult rat, behaviors displayed only during maturity, are dependent on adequate "programming" by sex steroids during a critical perinatal period, which is the time of birth ± 5 days. Recent data suggesting that human sexual behavior differences may also be hormonally mediated comes from the studies of Van Look and colleagues [11], who have shown that the release of sex-specific hormones (in particular, leutinizing hormone) by the endocrine system in response to treatment with female estrogen hormones is a predetermined sex-specific pattern in normal men and women.

**Sexual Dimorphic Behavior as Influenced by
Genetic Abnormalities**

To consider some of the evidence of hormonal control of behavior in humans, we now move to the three genetic disorders detailed earlier, each of which has effects on sexual development through hormonal interaction, with sexual behavior often being modified. Our goal is to show you how this type of medical-sociological approach to the study of interactions between genetics and environment can serve as a model for the study of numerous other nature-nurture relationships, and help to elucidate their relative roles in the determination of behavior in the adult human. These exemplary disorders and their essential biological relationships can be summarized as follows:

I. Testicular   genetic (chromo-   = gonadal sex
    feminization   somal sex
                                                ≠ morphological sex

                                                ≠ sex of rearing and   =
                                                  gender identity

II. Adrenogenital syndrome   genetic sex   = gonadal sex
(in 46,XX indi-
viduals):   = or ≠ morphological
sex

                                                 = or ≠ sex of rearing   =
and gender
identity

III. XYY Syndrome   genetic sex   = gonadal sex

                                               = morphological sex

                                               = sex of rearing and
gender identity

Recall that the syndrome of testicular feminization (TF, or androgen insensitivity) is an inherited problem of testosterone (androgen) nonresponsiveness in which chromosomal males fail to masculinize because they lack the receptor necessary at the cell level to utilize the hormone. Therefore, though such individuals have testes and produce androgens, the external genitalia remain completely feminine. This is a situation where genetic sex is male, but all other aspects of sexual identity are female. Effectively, these individuals are exposed to no androgens at all, so that their minor estrogen levels allow feminization at puberty and after. However, the HY antigen levels are those of a normal male, as the Y chromosome is indeed present. Individuals with the TF syndrome are raised as girls, with no parental ambiguity or early knowledge of the problem in the great majority of cases. They generally come to attention at puberty, when there are no spontaneous menses, or occasionally earlier if testes are detected in inguinal hernias. After the diagnosis is made on the basis of clinical and chromosome studies, the usual procedure is to explain to the patient that her internal female structures are incompletely developed, and that as a result she will not be able to bear her own children. No treatment is needed before puberty, but afterward testicular elements must be surgically removed to prevent malignant change, and female hormones are needed to ensure feminine characteristics.

This is one of the rare situations in genetic counseling in which the "truth" of opposite biological sex is kept from the patients. With appropriate management, these individuals can lead normal female lives with appropriate sexual function, except for their sterility. An individual with the TF syndrome, then, is unambiguously female from birth, and generally has no problems developing a female gender identity and role. The question to be asked in this situation is what might be the possible effects of the presence of the Y chromosome and the HY antigen of gender identity development and behavior, in the absence of the effects of adrenal and testicular androgens. Another interesting question as yet unanswered at this point is whether the lack of androgenic influences might have negative effects on some aspects of sexual function, such as sexual response. Though there is no clear evidence for this postulate, we know of several individuals with testicular feminization who are reportedly somewhat sexually promiscuous. One wonders if this could be connected to a lack of sexual satisfaction, since it is known that androgens play a very important role in at least male sexual response.

Because of the small number of patients known who have been diagnosed early and followed into adulthood with proper management, there are few good psychosexual studies to date of XY individuals with androgen insensitivity. However, Money and colleagues began in 1968 to report on such women. Ten were studied in the initial investigations, the findings of which were confirmed with four other patients later [12, 13]. However, before discussing some specific studies of sexual dimorphic behavior, as we will for both testicular feminization and the adrenogenital syndrome, we should divert to mention the types of characterstics generally studied, and the techniques employed. Several behaviors considered to be sexually dimorphic ones, from animal studies in many species as well as of normal human adults, have been postulated to result partially from the effects of prenatal androgenization of the brain. Three of these categories—intense physical energy expenditure, parental rehearsing in childhood, and the choice of peer contacts and group interactions—apply to nonhuman primates as well as to humans, whereas the other two often cited as classic examples of sexual dimorphic behavior are exclusively human considerations: gender-role labeling and attention to appearance [14]. These are the five types of sexual dimorphic behavior that have been the focus of most examinations of the role of prenatal sex hormones on gender identity and role. Because of the rarity of informative cases and the varying ages of detection of patients, investigative techniques generally have to depend on interviews with patients and family members at one or several ages.

All ten of Money's original subjects were diagnosed in late childhood or early puberty. Mean age of learning of their sterility was 15 years (range 9-19). Patients were followed for at least 6 years after diagnosis; several psychological interviews were conducted with patients, and with their family members. All ten had normal female sexual phenotype and function, and four were married at the time of study. This investigation concluded that patients with the androgen insensitivity syndrome are exclusively feminine, in both their gender role and identity. When questioned about marriage and maternalism these women generally preferred being a wife with no outside career (80 percent), indicated that they enjoyed homemaking (70 percent), had fantasies concerning raising a family (100 percent), and as children had played primarily with dolls and other girls' toys (80 percent). Though most were well-resigned to the fact that they could not bear children, they showed a strong interest in infant care (see table 7-2). Those who had adopted children (two) proved to be good mothers with a strong sense of mothering. When asked to rate themselves on affection, 80 percent felt that they were moderate or high on the scale; their husbands agreed with these assessments and all reported satisfaction with their marriages. On the criteria of sex and eroticism, the testicular feminization patients stood out as strongly feminine, especially when compared to those women with the adrenogenital syndrome (to be discussed later). The majority (90 percent) of TF individuals rated themselves as fully content with the female role; the tenth was a teenage girl with multiple psychological problems in regard to family and religion as well as gender role. The authors therefore concluded as follows: "These various percentages clearly tell a story of women whose genetic status as males was utterly irrelevant to their psychosocial status as women, as also was the histology of their gonads."[13].

The second disorder to discuss is one in which genetic sex and gonadal sex may be quite different from morphological sex, and therefore sex of rearing. In the recessively inherited adrenogenital syndrome, an enzymatic defect leads to excessive androgenization beginning in fetal life. If the fetus is female, there will be female internal structures, but masculinizaton of the external genitalia of varying degrees, as well presumably as masculinization of the fetal brain prenatally. Since the degree of external male development may be such that the individuals are proclaimed male at birth and raised as such, there is understandably often ambiguity about the diagnosis and the sex of the infants. Though this is a poor start for the family, who must begin to program their behavior toward an infant of one sex or the other, if the case is well handled the counsel and necessary corrective surgery will be completed early. The decision of sex of rearing is usually based primarily on the probability that a somewhat masculinized individual can function in society and sexually as a male, or whether feminization could allow a more satisfactory life.

Therefore, in adrenogenital syndrome we have a situation where a genetic female, with ovaries and internal female structures, may be raised as either male or female. Androgen levels are high in utero and until corrective enzyme therapy is initiated. However, there is no Y chromosome and no HY antigen present in females affected with this disorder. In this regard, assuming early proper management and decision as to the sex of rearing, what can this disorder tell us about the effects of fetal androgens on subsequent behavior?

Table 7-2
Marriage and Maternalism in Ten Patients with Testicular Feminizing Syndrome

Class of Information	Number of Patients
A. Homemaking and marital expectations	
1. Priority of marriage vs. career	
a. equally important	8
b. wife with no outside job	2
2. Domestic activities	
a. indifferent	2
b. enjoys taking care of house	7
B. Pregnancy and delivery	
1. Dreams and fantasies	
a. Yes	3
b. No	7
2. Attitude toward physical experience of pregnancy and delivery	
a. negative	6
b. noncommittal	1
c. positive and genuine	3
C. Attitude toward raising a family	
1. Dreams and fantasies	
a. Yes	10
2. Anticipation of wife and mother role in childhood play	
a. mainly with dolls, "girl's toys"	8
b. mainly with trucks, guns, etc.	1
3. Preference of newborn versus other age levels in children	
a. negative	6
b. positive and genuine	3
4. Outlook on infant care	
a. negative	3
b. noncommittal	1
c. positive and genuine	6
5. Patient's degree of affectionate behavior (personal estimate)	
a. very affectionate	5
b. average	3
c. little affectionate behavior	2

Source: J. Money; A.A. Ehrhardt; and D.N. Masica, Fetal feminization induced by androgen insensitivity in the Testicular Feminization syndrome: Effect on marriage and maternalism, *Johns Hopkins Medical Journal* 123 (September 1968):109. Reprinted with permission.

There have been several systematic studies of individuals with the adrenogenital syndrome. One of those, conducted by Money and Dalery [15, 16], examined behavioral characteristics of seven 46,XX individuals with this endocrine disorder, three of whom had been raised as boys and four as girls. These individuals, who ranged in age from 5 to 28 years at the time of study, were examined by medical records and personal interviews. A standard schedule of behavioral topics was followed, which included childhood play habits and interests, with rehearsal of adult parenting roles, romantic interests in adolescence and adulthood and sexual practices and fantasies (see table 7-3). According to these investigations, the authors concluded that those individuals raised as males, though genetically female, were totally male in their habits and attitudes, with boyish play habits and interests, intense athletic tendencies, male peer associations, and male sexual interests and gender role identity. For those AGS patients raised as females, however, traditional adaptations were not so good. In childhood they expressed early preference for male companions, and "tomboyish" activities and competitiveness predominated. They had little interest in personal appearance, and preferred functional, utilitarian clothing. Though most were too young at the time of study to assess satisfaction with adult sexual function, all were noted to be "late bloomers" in adolescence. The authors therefore concluded that the early androgenization apparently had an effect on later sexual dimorphic behavior. They suggest that in cases of extreme virilization the option of raising 46,XX genetic females with AGS as males should be seriously considered, as the transition through adolescence to adult psychosexual maturity may be easier for the individual raised as male.

A study by Ehrhardt and Baker [17] compared AGS individuals with their normal siblings, in order to minimize the effects of social environment on gender identity development. They studied seventeen female and ten male patients, in whom genetic sex and sex of rearing were concordant, and who were evaluated in childhood and early adolscence. They utilized the classic criteria of sexually dimorphic behavior in childhood also, in regard to (1) activity and aggression, (2) rehearsal of marriage and parenthood, and (3) gender role and clothing preferences. Their conclusions strengthened those of earlier studies, that the boys had 100 percent traditional male habits, with the only significant difference from their male siblings being a more intense energy expenditure in outdoor play and sports. However, when AGS females were compared with unaffected female siblings and mothers, the trend emerged that these girls tended to be long-term tomboys, who had significantly less interest than their same-sex family members in traditional female play, and in the rehearsal of marriage and mothering. These authors too suggested that prenatal androgen is an important factor in the development of temperamental and functional behavior differences between the sexes.

Table 7-3
Summary of Studies of Congenitally Masculinized 46,XX Individuals

	Individuals Reared as Males (N = 3)	Individuals Reared as Females (N = 4)
Childhood rehearsal of sex-coded roles (play habits, domestic activity)	"Normal" boyish play habits and interests	"Tomboyish" activities predominate (traditional male toys and interests)
	Lack of interest in infant caretaking	
Physical energy expenditure	Intense athletic interests	"Tomboyish" energy expenditure and competitiveness
Peer association preference	Male	Preference for males
Cosmetic interests	Unremarkable for males	Functional, utilitarian clothing; diminished interest in appearance
Satisfaction with own sex	Good	Reasonably good
Social mixing/acceptance	Good	Good
Sexual practices—early childhood	Normal investigative play	Normal investigative play
Romantic interests	Normal since early adolescence	Too young to assess—little current interest (generally "late bloomers")
Adult sexual function and satisfaction	Normal heterosexual	Too young to assess

Source: J. Money and J. Dalery, Introgenic homosexuality: Gender identity in seven 46,XX chromosomal females with Hyperadrenocortical Hermaphroditism born with a penis, three reared as boys, four reared as girls, *Journal of Homosexuality* 1(Summer 1976):357–371. Reprinted with permission.

If these studies of adrenogenital syndrome are accepted as evidence of the effects of androgens on sexual development of the brain, there is at least one confusing factor that can be clarified by discussion of another developmental sexual problem. Some of the AGS females have had initiation of hormone therapy only after a few months have passed, raising the question of whether the important androgenization is indeed prenatal in origin, or whether it might be postnatal in effect. In this regard another study by Money and Ehrhardt [16] is enlightening. When these authors examined psychosexual behavior and energy expenditure in a group of genetically normal females whose mothers took drugs with androgenizing side effects during their pregnancies, they found the same patterns in these girls androgenized only prenatally, as in the AGS females who had been androgenized both prenatally, and to some extent at least, postnatally.

The third disorder we will investigate is the XYY syndrome, a relatively common, sporadic chromosomal disorder (1/600–1/800 live births) [18]. Though not yet adequately understood, we know that the phenotypic effects of the extra Y chromosome are often the following: stature is increased and puberty often aggravated, but intelligence, behavior, and fertility are adversely affected less often. For our topic here, however, it is important to note that individuals are unquestionably masculine at birth, and if they attract medical or psychological attention at all, it is rarely before adolescence. Though there is no sexual ambiguity physically or psychologically, there are excesses of both HY antigen and possibly of androgens, apparently dictated by the extra Y chromosome. The question to be considered, then, is what effects these masculinizing factors may have on the individual in terms of behavioral adjustment.

It should be remembered that individuals with the XYY anomaly often lead normal lives in society. To illustrate this disorder we will present the case of Danny, a young man whose parents consulted us when he was seven years old. As he was rather hyperactive and clumsy, they were concerned that these might be symptoms of the Duchenne muscular dystrophy that affected some cousins.

After discussing the family history it was evident that the child was not at risk genetically for this inherited type of muscular disease. Nonetheless, he exhibited some of the features that have come to be associated with the XYY phenotype: excessively tall stature, short attention span, near hyperactivity, rather aggressive demeanor, and poor interpersonal relations. His school performance thus far had been borderline. A chromosome study demonstrated an extra Y. After careful counseling of the parents in regard to this disorder, we recommended that they consider psychological intervention starting at an early age in an attempt to alleviate potential adjustment problems.

Studies of the behavioral implications of the XYY karyotype have been fraught with controversy, due to a regrettable early publicized association with criminal behavior [19, 20]. The first few studies were either of individual cases or of inmates in mental or penal institutions. Today, however, after several well-planned studies have been completed, we are left with the evidence that the XYY karyotype carries some propensity for behavioral abnormalities. Table 7-4 summarizes the results of several studies of the XYY in various populations; though the statistics differ, we find that the frequency of the XYY in mental or mental-penal institutions is apparently 4–20 times higher than that in the male population at birth [7, 8]. Behavioral problems noted even in noninstitutionalized adults include lessened adaptive capacity, poor self-control and impulsivity, with emotional outbursts [21, 22]. Criminality is apparently somewhat increased. Several thorough studies that took into account family histories and socioeconomic background have demonstrated that the average age of first offense in the criminal population is younger for the XYY criminal than for his matched XY peer. In addition, XYY criminals come from a broad socioeconomic spectrum, while the association of lower socioeconomic status with XY criminals is quite strong. On the other hand, XY criminals often have siblings with criminal records, whereas the incidence of convicted siblings of the XYY males has been lower (10 times more for the siblings of XYs) [5, 23]. These comparisons favor the hypothesis that it is the presence of the extra Y itself that predisposes to aberrant behavior rather than familial or socioeconomic factors. However, it is important to note that these studies dispel a popular myth about the XYY male; the nature of their crimes is apparently not violent, as originally reported; rather, the offenses are those against property, etc.

One of the better studies of XYY males is that of Owen and colleagues [24]. Since tall stature is a relatively constant feature of the XYY male, these investigators studied 4,139 tall adult males (184 cm or more) from a total of 31,436 consecutively born in Denmark during a four-year period. In this tall male population (who comprised 15.9 percent of all male births) they found 12 XYY males (2.9 percent of those studied). The primary goal after identifying these individuals was to see if there was any increase in criminal behavior. They compared the tall XYYs with tall XYs, matched for socioeconomic status, intelligence (from draft board screening), and level of education. Of the 12 XYY males, 5 (42 percent) had at least one felony account, as compared to 9.3 percent of the XY controls. However, there was only one habitual criminal (whose approximately fifty offenses had been mostly for thefts of motor vehicles), and the crimes for which the XYYs had been convicted were not particularly those of acts of aggression against other people. There were no differences between XYs and XYYs in the rate of violent crime. To further elucidate the causes of criminality, men of both groups who had committed crimes were closely matched for variables of parental social class as well as those mentioned above. Statistical compar-

Table 7-4
XYY Frequencies

In population	1:1000 (or greater)	
In sixteen penal institutions	4:1000	(26/5,805)[a]
In nine mental institutions	3:1000	(8/2,526)[a]
In twenty mental-penal institutions	20:1000	(72/3,813)[a]

[a]Hook, E.B. *Science* 179:139 (1973); *Lancet* 1:98 (1975).

isons of probability based on these variables demonstrated that such factors accounted for only half of the increased incidence of crime in XYY males over the XYs. In another aspect of their study, these authors also found that XYYs from the general tall population had significantly depressed IQ scores, though they seemed to fall in the low normal range. However, various WAIS subscales and factor scores from more complete testing failed to show any specific cognitive pattern associated with the XYY condition.

In summary, then, the XYY syndrome affects individuals of male genetic sex and gender identity, who have some special physical and behavioral predispositions, including average lower intelligence, impaired social adaptability, impulsive behavior, and a higher rate of crimes against society. Though masculinity-determining factors are consistent with genetic sex, both HY antigen and perhaps also male hormones [5] are augmented, raising the question of their potential effects on the particular behavioral anomalies discussed.

In conclusion, these "experiments of nature" have shown us that many biological factors play a role in the development and determination of normal dimorphic sexual function and behavior. As these biological elements become better defined and understood, they indicate more and more strongly a truth that has been underestimated in the recent past. Though environmental, educational, and other biographical elements have a known importance in such development, it is increasingly clear that genetic, and more specifically hormonal, influences acting already during embryo-fetal life are very important and at times predominant in the production of some human behaviors and functions. Recent investigations of human normal and pathological states in this regard have amplified the already existing evidence from various animal species that some behavioral imprinting of the brain occurs prenatally. The natural models that we have discussed in this chapter—the testicular feminization, adrenogenital, and XYY syndromes—provide human evidence for such hormonal effects. Possibly the most important prenatal genetic influence on behavior can be attributed to the androgenic hormones. As aptly expressed by Ohno [25] ". . . Human genes awakened by androgens have been whispering to the ears of the sexually mature man . . . ."

Postscript

Since our presentation of a paper in April 1979 that resulted in this chapter, a report has been published that, though somewhat controversial, strengthens the concept of prenatal androgenic influences on gender identity and role. Imperato-McGinley and colleagues [26] conducted their studies in two inbred villages in which there is a relatively high incidence of a recessive genetic disorder involving the metabolism of testosterone (deficiency of 5 alpha-reductase). The result of this metabolic defect is a lack of the androgen dihydrotestosterone, which is needed for the developmental masculinization of fetal genitalia. As genetic males with this disorder do not masculinize adequately in utero, genital ambiguity at birth is such that most individuals are raised as females. However, since internal genitalia are male and the testes produce testosterone, some androgenic influence is present from birth, and these individuals masculinize strongly at puberty. Of eighteen affected individuals unambiguously raised as girls, seventeen to eighteen decided to change to male gender identity and sixteen to eighteen to a male gender role during or after pubertal masculinization. The authors therefore concluded that exposure of the brain to normal levels of testosterone, prenatally and at puberty, contributes substantially to the formation of gender identity. Indeed, in such a "laissez-faire" societal attitude as exists in these villages, these early androgenic effects on sexual gender identity apparently predominated over the influences of rearing as females.

Other new evidence in regard to factors determining masculinization comes from a recent report by Eicher and colleagues [27] of HY antigen levels in transsexuals. According to their study transsexuals, who have a psychological rather than a morphological identification with the opposite sex, may have HY antigen levels incongruent with their morphological sex. They reported that of seven male-to-female transsexuals, four were HY antigen negative (normal female pattern), two positive, and one weakly positive. Five of seven female-to-male transsexuals were HY antigen positive, one weakly positive, and one negative. If these findings are verified, they might indicate that HY antigen plays a role not only in morphological sexual differentiation, but in the later development of gender identity as well.

References

1. Wilson, J.D., and Goldstein, J.L. 1975. Classification of hereditary disorders of sexual development in genetic forms of hypogonadism. *Birth Defects, Orig. Artic. Ser.* 11(4):1.
2. Grumback, M. 1979. Genetic mechanisms of sexual development. Pp.

33–74 in *Genetic Mechanisms of Sexual Development.* H.L. Vallet and I.H. Porter, eds. Academic Press, New York.

3. Federman, D.D. 1967. *Abnormal Sexual Development: A Genetic and Endocrine Approach to Differential Diagnosis.* W.B. Saunders, Philadelphia.

4. Short, R.V. 1979. Sex determination and differentiation. *Med. Bull.* 35(2):121.

5. Simpson, J.L. 1976. *Disorders of Sexual Differentiation: Etiology and Clinical Delineation.* Academic Press, New York.

6. Franks, R.C.; Bunting, K.W.; and Engel, E. 1967. Male pseudohermaphroditism with XYY sex chromosomes. *J. Clin. Endo. Met.* 27(11): 1623.

7. Hook, E.B. 1973. Behavioral implications of XYY. *Science* 179:139.

8. ———. 1975. Rates of XYY genotype in penal and mental settings. *Lancet* 1:98.

9. Short, R.V. 1972. Pp. 122–142 in *Endocrinologie Sexuelle de la Periode Perinatale.* M.G. Forest and J. Bertrand, eds. INSERM, Paris.

10. Shapiro, B.H., and Goldman, A.S. 1979. New thoughts on sexual differentiation of the brain. Pp. 221–251 in *Genetic Mechanisms of Sexual Development* H.L. Vallet and I.H. Porter, eds. Academic Press, New York.

11. Van Look, P.F.; Hunter, W.M.; Corker, C.S.; and Baird, D.T. 1977. Failure of positive feedback in normal men and subjects with Testicular Feminization. *Clin. Endocrin.* 7:353.

12. Money, J.; Ehrhardt, A.A.; and Masica, D.N. 1968. Fetal feminization induced by androgen insensitivity in the Testicular Feminization syndrome: Effect on marriage and maternalism. *Johns Hopkins Med. J.* 123:105.

13. Money, J. 1973. Intersexual problems. *Clin. Obstet. Gynecol.* 16:169.

14. Ehrhardt, A.A., and Meyer-Bahlburg, H.F.L. 1979. Psychosexual development: An examination of the role of prenatal hormones. Pp. 41–57 in *Sex Hormones and Behavior.* R. Perler and J. Whelan, eds. Ciba Foundation Symposium 62. Elsevier/North Holland, Amsterdam.

15. Money, J., and Dalery, J. 1976. Introgenic homosexuality: Gender identity in seven 46,XX chomosomal females with Hyperadrenocortical Hermaphroditism born with a penis, three reared as boys, four reared as girls. *J. Homosex.* 1(4):357.

16. Money, J., and Ehrhardt, A. 1972. *Man and Woman, Boy and Girl.* Johns Hopkins Press, Baltimore.

17. Ehrhardt, A., and Baker, S.W. 1974. Fetal androgen, human central nervous system differentiation, and behavior sex differences. Pp. 55–76 in *Sex Differences in Behavior.* R.C. Friedman et al., eds. Wiley, New York.

18. Hamerton, J.L.; Canning, N.; Ray, M.; and Smith, S. 1975. A cyto-genetic survey of 14,069 newborn infants. I. Incidence of chromosomal abnormalities. *Clin. Genet.* 8:223.
19. Jacobs, P.A.; Brunton, M.; Melville, M.; Brittain, R.P.; and McCle-mont, W.F. 1965. Aggressive behavior, mental subnormality and the XYY male. *Nature* (London) 208:1351.
20. Engel, E. 1972. The making of an XYY. *Am. J. Ment. Defic.* 77(2): 123.
21. Noel, B. The XYY syndrome: Reality or myth? *Clin. Genet.* 5:387.
22. Walzer, S.; Gerald, P.S.; and Shah, S.A. 1978. The XYY genotype. *Ann. Rev. Med.* 29, 563.
23. Price, W.H.; Strong, J.A.; Whatmore, P.V.; and McClemont, W.F. 1966. Criminal patients with XYY sex-chromosome complement. *Lancet* 1:565.
24. Owen, D. 1979. Psychological studies in XYY men. Pp. 465–471 in *Genetic Mechanisms of Sexual Development.* H.L. Vallet and I.H. Porter, eds. Academic Press, New York.
25. Ohno, S. *Major Sex-Determining Genes.* Springer Verlag, New York.
26. Imperto-McGinley, J.; Peterson, R.E.; Goutier, T.; and Sturla, E. 1979. Androgens and the evolution of male-gender identity among male pseudohermaphrodites with 5 α-reductase deficiency. *New Engl. J. Med.* 300(22):1233.

8 Biology, Experience, and Sex-Dimorphic Behaviors

Jacquelynne Eccles Parsons

The question of the existence and origin of sex differences has been debated by philosophers and scientists for centuries. Much of the debate has centered around the relative importance of biological versus experiential influences. At one extreme, it has been argued that men and women are destined by biology to play quite different roles in society and to have quite distinct personalities (see Freud 1965). At the other extreme, it has been argued that sex roles in modern-day society result totally from markedly different socialization experiences for boys and girls; biology is assumed to play a minimal role in the maintenance of sex roles. Most scientific investigators today do not take a simple either-or position concerning the determinants of sex differences. Instead, human development is seen as the result of the dynamic interaction between an individual's biological makeup and experiences with the environment. The crux of the debate lies in disagreement over the exact nature of the interactions between experience and biology in shaping the sex-dimorphic patterns associated with some social behaviors.

In considering the interactive role of biological and experiential processes, several important issues arise. First, we must identify those sex-dimorphic behaviors that are influenced to some degree by biological processes. It is safe to assume that all behavior is influenced by socialization and experience. Man is just too adaptable for this not to be true. It is also very likely that some behaviors are shaped totally by experiential factors. The question becomes, Which of the remaining behaviors are shaped at least to some degree by biological processes? Typically, social scientists have relied on four sources of information in deciding which sex-dimorphic behaviors these might be: (a) demonstrations of an association between hormonal and behavioral variations, (b) behavioral patterns among infants or very young children, (c) cross-cultural universals, and (d) cross-species consistency, especially among higher primates. While a clear result in any one of these categories suggests the importance of biological influences, congruent findings from two or more categories provide much stronger evidence of a biologically mediated mechanism. Taking this more conservative criterion, it now seems likely that biological processes are involved in the following sex-dimorphic behavior clusters: aggression and/or activity level; a set of limited cognitive skills associated with spatial visualizations and

137

perhaps mathematical reasoning and verbal skills as well; and parenting. Each of these will be discussed in more detail later.

Having identified plausible candidate behaviors, the next question becomes, How are the biological effects mediated? To say something has a biological basis does not, in and of itself, tell us much. It simply narrows the search of possible causal determinants. Real progress depends on the identification of the specific biological mechanisms responsible. In addition, the nature of the interaction between specific biological and experiential processes in shaping behavioral development must be delineated. Biological processes do not unfold in a cultural experiential vacuum. Likewise, experience does not accumulate in a biologically neutral organism. The interaction between these two forces is undoubtedly quite complex and extremely varied in its specific details. Nevertheless, a complete understanding of the origins and development of sex-role dimorphic behavioral systems is dependent on our knowledge of both the biological mechanisms themselves and the nature of their interaction with experiential forces.

In seeking this understanding, two additional issues need to be considered. First, it should be noted that the distinction between biological and experiential causes is rarely a clear issue. The empirical data are generally ambiguous enough to allow room for interpretations based more on the scientist's theoretical perspective than on the actual data. Key to these interpretations is the individual scientist's assumptions regarding the behavioral phenomena to be explained, the presumed mechanisms of the biological effect, and the malleability of this effect. Take the impact of anatomical differences on development as an example. In deeming anatomy to be destiny, Freud was suggesting that a child's anatomical structure, which is biologically determined, has an *inevitable* and *irreversible* effect on the child's personality development that is *independent* of any differential treatment from socialization agents. This stance has been classified within the domain of biological influences on behavioral development because it stresses an *inevitable* effect of anatomical features that originates *inside* the individual.

In contrast, a number of investigators have focused on the effect of the child's anatomical sex on caregiver behavior. For example, several studies have demonstrated that people respond differently to the same baby depending on whether they think the baby is a boy or a girl (for example, Condry and Condry 1976). While these studies suggest that one causal source of parents' behaviors is a child's anatomy, this work is generally cited as evidence supporting an experiential explanation of the origin of sex-role dimorphism. The link of anatomy to behavioral development is *not* assumed to be direct, internally generated, or inevitable; it is assumed that caregiver responses could be changed if the meaning of anatomical differences were changed and that the impact of caregiver responses can be modified by subsequent experiences. Thus, while these theorists acknowledge that anatomy is a biological event of some note, they assume that it exerts its influence primarily through social processes.

Second, one has to be aware of the fact that there is a wide range of effects that might reflect biological processes. For example, hormones have a direct effect on prenatal morphological development leading to anatomical sex dimorphism (see Money and Ehrhardt 1972; Wilson, George, and Griffin 1981). Hormones also have a direct effect on the development of the brain such that exposure to specific prenatal steroids programs subsequent brain activity including cyclicity of gonadal control and postpubertal responsivity to the various steroids (see McEwen 1981). Further, studies with primates clearly suggest that the effects of prenatal exposure to various steroids on the brain include behavioral dimorphism as well as anatomical and neural dimorphism (see Bardin and Catterall 1981; McEwen 1981; Money and Ehrhardt 1972; Reinisch 1981). For example, exposure to androgens has been found to increase the frequency of rough-and-tumble play in primates. These examples illustrate a direct link between a specific biological process and a sex-dimorphic consequence.

At a more indirect level, boys and girls may differ not because of the sex-linked hormonal effect itself but because of interactions between gender and other biological processes that themselves have a direct effect on behavior and experience. Consider for example maturation rate, a biological process that influences both behavior and experiences. Girls on the average mature more rapidly than boys. They are born neurologically more mature, pass many of the developmental milestones earlier, and reach sexual maturity sooner than boys (see Frieze et al. 1978). Because boys and girls differ in their maturational rates, they may develop different skills, thus eliciting different responses from their social environment. For example, Sherman (1971) has suggested that the early advantage girls have in language could predispose them to rely on verbal modes of problem solving rather than to acquire both verbal and nonverbal problem-solving skills. This, in turn, could account for the fact that girls do less well than boys on tasks requiring spatial visualization. Waber (1979) has suggested an alternative explanation for the sex difference in spatial visualization that also relies on the fact that girls reach sexual maturity earlier than boys. She suggests that brain lateralization (the degree to which each hemisphere of one's brain specializes in certain functions) proceeds until one reaches puberty, at which point brain lateralization, like growth, slows dramatically. Because boys mature later, their brains are more lateralized. Finally, Waber argues that greater lateralization facilitates spatial visualization and consequently, delayed puberty is responsible, in part, for males doing better on the average than females on spatial-visualization tasks. These examples indicate that some sex-dimorphic patterns may be mediated by the interaction of sex and other biological processes such as maturation rate rather than by biological processes linked more directly to sex such as the hormonal effects discussed in the previous paragraph.

Even more indirectly, biological processes may affect some factor that is correlated with sex (like size) and is *assumed* by parents to be related to other factors (like fragility). Consequently, parents may respond differently to boys and girls and thus socialize, unnecessarily, consistent patterns of sex-dimorphic behavior. For example, males are born larger and remain somewhat larger than females. Size and muscle mass may be linked phenomenologically to perceived fragility. In turn, boys may be assumed to be tough and because, in fact, they are born bigger and have stronger neck muscles they may be treated less gingerly than girls. As a consequence boys may develop more active play patterns (see Maccoby and Jacklin 1974).

It can be seen from these examples that determination of what constitutes a biological effect is a complex issue. Biological processes can impinge on sex-role dimorphism directly (females have babies while males do not); males have penises while females do not; males have higher levels of testosterone; females have higher levels of estrogens and progesterone), or quite indirectly (through maturational rates, body size, or morbidity rates). Further, no matter how the biological processes are manifest, their influences on behavior are undoubtedly mediated by their interaction with experiential forces. It is to this issue that I now turn.

As was stressed in the preceding paragraph, the delineation of biological effects independent of consideration of experiential effects is impossible. Biological processes do not unfold in a cultural, experiential vacuum. Likewise, the delineation of experiential effects independent of a consideration of biological processes is futile, if not impossible. Experience does not accumulate in a biologically neutral organism. In addition, neither of these processes (biological and experiential) take place in a sociohistorically neutral context. It is the interactions of all of these processes (biological, psychological, and sociohistorical) that determine behavior.

Peterson (1980) has recently outlined a dialectical model of the interactive effects of biology and experience based on the thinking of Sameroff (1977) and Riegel (1976). The model makes the following assumptions. First, individuals continue to grow and change throughout their life spans. Second, this growth is determined by the interplay of biological, psychological, sociocultural, and historical processes. Third, the interactive nature of development is itself not static but is shifting and accumulating across time. For example, at some points in one's life span, biological processes may have greater influence than at other times, for example, during the prenatal and pubertal periods or while one is pregnant. Similarly, the nature of the interaction between experience and biology may vary across one's life time. For example, maturational rate may have little influence on the effects of experience among twenty-year-olds but may have a direct and powerful effect on the entire course of one's early adolescent experiences. Furthermore, a specific form of biological or experiential interaction can have a lasting effect on all subsequent development of one person and little or no effect on another person. For example, consider a girl who matures very early. Not only will this biological event affect basic processes such as brain lateralization; it will also affect social events at a very critical developmental time. If attractive, she may be drawn into an early social-dating pattern that, in turn, may distract her from her studies. Thus, her rate of maturation can affect both her career possibilities and the spatial-visual training she receives during high school. Together these forces could shape the direction of her future life as well as the social experiences she has while an adolescent.

As can be seen in the previous example, it is not only the relative importance of biological and social processes that one must consider; the very form of the interaction between them may change as an individual grows and develops. Consequently we can not expect an easy answer to the question of the origin of sex-role dimorphism. It is necessary to specify not only the particular behavior but also both the developmental age of the individuals being considered, and the sociocultural environment in which these individuals are developing. Such an analysis will not only help us to describe the interaction of biological and experiential forces at one point in time; even more interestingly, such an analysis will also lead us to an investigation of the nature of the interactive processes themselves rather than to a static analysis of the differential causes of temporally fixed behavioral events.

Before turning to a discussion of specific patterns of behavior I must address the final issue—the malleability of biological effects. Biological processes are often assumed to be stable, inevitable, and relatively immutable while social and psychological processes are assumed to be more unstable, variable, and arbitrary. Both these assumptions are now being questioned, especially as we learn more about specific biological mechanisms and about their interaction with social forces. Indeed, identifying the specific biological mechanisms underlying a particular behavior may provide the means of behavioral modification and cultural change rather than sealing our "fate." For example, tooth decay is determined in part by soft enamel, which is an inheritable characteristic. Soft enamel can be eliminated, however, by providing fluoride to developing fetuses and young children. In this instance, then, an environmental manipulation can override a biologically based individual difference. Similarly, while increasing evidence points to the role of sex-dimorphic brain lateralization patterns in sex-differentiated spatial-visualization skills (Wittig and Peterson 1979), appropriate training can largely eliminate the sex differences in performance (Connor, Serbin, and Schackman 1977). Thus, to conclude that some characteristic or behavioral system has a biological substratum is *not* to say that it is immutable to exogenous influences.

Having introduced the complexities of the relation between biology and experience in shaping sex-dimorphic patterns of behavior, I will now turn to a more specific discussion. As noted earlier, there is some consensus now that biological processes are implicated in certain sex-dimorphic behavior patterns. In particular, sex differences in certain limited cognitive skills, in aggression, and in parenting behaviors occur with such regularity that it is highly probable that biological processes are involved. The remainder of the chapter will be devoted to a discussion of these behavior patterns and the current state of our understanding of the forces shaping these patterns.

Sex Differences in Cognitive Skills

Three areas of cognitive functioning are commonly cited (see Maccoby and Jacklin 1974; Wittig and Petersen 1979) as revealing fairly consistent patterns of sex differences. These are verbal skills, quantitative skills, and spatial visualization. The findings with regard to verbal skills (a) are equivocal, depending on the particular measures of verbal skills used; (b) seem to be limited primarily to area of verbal fluency; and (c) are either not present or are not very large among older adolescents and adults (cf. Frieze et al. 1978). Furthermore, no consistent theories of possible biological mediators have emerged with the exception of early brain lateralization and maturation rates. These theories will be discussed in more detail in connection with spatial skills.

More extensive research has focused on the sex differences in spatial-visualization and quantitative skills. In both areas the patterns of results are fairly consistent both within our culture and cross-culturally. Males do better than females in these skills after puberty. Since these differences are often linked to each other, and since the sex difference in spatial skills is repeatedly nominated as the cause of the sex difference in tests of quantitative reasoning, the evidence for each will be discussed separately; then the possible link between them and the possible biological mediators of both differences will be discussed together.

Quantitative Skills

The pattern of sex differences in mathematics achievement is fairly consistent across studies using a variety of achievement tests. High-school boys usually do somewhat better than girls on tests of mathematical reasoning (primarily solving word problems); boys and girls do about the same on tests of algebra and basic mathematical knowledge; and girls occasionally outperform boys on tests of computation skills. The differences favoring boys, however, do not emerge with any consistency prior to the tenth grade, are typically not very large, and even in the advanced high school groups are not found universally (see Fox, Brody, and Tobin 1980; Maccoby and Jacklin 1974; and Wittig and Petersen 1979 for reveiw of studies prior to 1975. More recent work includes Burnett, Lane and Dratt 1979; Fennema and Sherman 1977, 1978; Schratz 1978; Sherman 1980a, in press; Starr 1979; Armstrong 1980; Connor and Servin 1980; ETS 1979; Steel and Wise 1979). Several of the recent studies have used large national samples (for example, Armstrong 1980; ETS 1979; and Steel and Wise 1979). The pattern of results with these national samples is quite consistent and provides strong support for the conclusions reached above. Nonetheless, even these studies reveal some inconsistencies, and studies of smaller, more specialized samples yield an even more inconsistent picture. For example, while ETS (1979) replicated the typical sex differences on the Scholastic Aptitude Test (SAT) scores, the magnitude of this difference varied across ability levels, being most pronounced in the top 10 percent of the students. In addition the school grades of the test takers did not differ by sex. Similarly, Fennema and Sherman (1977) found the expected sex difference in achievement in only two of four high schools studied; and finally, Schratz (1978) found that the direction of the sex difference varies across different ethnic groups.

There is one recent study that runs counter to the developmental pattern commonly reported. Benbow and Stanley (1980) have found a very consistent pattern of sex differences in mathematical achievement among highly gifted seventh-grade participants in the Johns Hopkins Study of Mathematically Gifted Youth. Year after year these boys outperform the girls by an average of 30 points (equivalent to 2 problems) on the Scholastic Aptitude Test for Mathematics. Why this might be the case is not yet known. But, it is interesting to note a consistency between this finding and the finding reported by ETS (1979). Sex differences on SAT-M are more marked among the most gifted high-school test takers. Apparently, sex differences also emerge earlier among the gifted.

In conclusion, adolescent males typically outperform adolescent females on tests of mathematical achievement. Furthermore, in most studies these differences exist, although to a lesser degree, even when one corrects for the number of mathematical courses taken (Armstrong 1980; Sherman 1980a; Starr 1979; Steel and Wise, 1979). But, even though the pattern of sex difference is fairly consistent, it is not inevitable; when it is found it is generally small; it is not apparent in the normal population prior to adolescence; and it is not typically reflected as a sex difference in course grades.

Spatial Skills

As was the case with mathematical achievement, the findings regarding sex differences in spatial skills are fairly consistent, though not universal, and do not emerge prior to the tenth grade. After junior high school boys begin to outperform girls on some measures of spatial skills (see Maccoby and Jacklin 1974; and Wittig and Petersen 1979 for reviews of earlier studies). Recent studies (including Burnett, Lane, and Dratt 1979; Connor, Schackman, and Serbin 1978; Fennema and Sherman 1977, 1978; Sherman, 1980*a,* Steel and Wise 1979) support the earlier conclusions but suggest that the magnitude of this effect varies depending on maturational timing (Waber 1979; Herbst and Petersen 1980), on body type (degree of masculinization; Petersen 1979), on personality characteristics associated with masculinity and femininity (Nash 1979), on previous experience with spatial activities (Connor, Schackman, and Serbin 1978; Sherman 1980*c*), on ethnic background, parental styles, and socioeconomic status (Fennema and Sherman 1977; Nash 1979; Schratz 1978; Gitelson 1980), and on the particular test given (Connor and Serbin 1980). In fact, in a recent national survey study of 3,240 junior and senior high-school students, thirteen-year-old girls were found to do better on a test of spatial skill than thirteen-year-old boys; twelfth-grade boys and girls did equally well (Armstrong 1980). Thus, as Connor and Serbin (1980) conclude, "junior and senior high school males . . . perform better than females on some visual-spatial measures, some of the time" (p. 36).

Relation of Spatial Skills to Quantitative Skills

The possibility that sex differences in spatial skills (supposedly biologically determined) mediate sex differences in mathematical achievement has become a popular hypotheses (see Burnett, Lane, and Dratt 1979; Hyde, Geiringer, and Yin 1975; Maccoby and Jacklin 1974; Sherman 1967; Wittig and Petersen 1979; Connor and Serbin 1980; Sherman 1980*c*). In assessing the hypothesis that the sex difference in spatial skills underlies the sex difference in mathematical achievement, three issues need to be discussed. First, is there a relation between spatial skills and mathematical achievement? Second, is this relation equivalent for both boys and girls? And third, does the sex difference in spatial skills mediate the sex difference in mathematical achievement?

With regard to the first question, several studies have demonstrated a strong positive correlation (*r,* ranging from .50 to .60) between spatial skills and a variety of measures of mathematical achievement test scores (Burnett, Lane, and Dratt 1979; Fennema and Sherman 1977, 1978; Sherman 1980*a;*

Armstrong 1980; Connor and Serbin 1980; Steel and Wise 1979). But it should be noted that verbal abilities also correlate quite highly with mathematical performance (Fennema and Sherman 1977, 1978; Sherman 1980*a;* Armstrong, 1980; Connor and Serbin 1980). In the most comprehensive study of the relation between spatial skills, verbal skills, and mathematical achievement, Connor and Serbin (1980) found a very inconsistent pattern of relations. While some measures of both verbal and spatial skills emerge as significant predictors of general mathematical achievement, not all measures of spatial skills correlated significantly with all measures of mathematical achievement, and the patterns of these relations varied among grade level and sex. Furthermore, when they factored their measures, the spatial skills scores factored together and independent of the factors containing all the measures of mathematical achievement.

Thus, it appears that the relation between spatial skills and mathematical achievement is not yet fully understood. While studies have yielded a fairly consistent positive relation between these two cognitive tasks, whether this relationship has the unique quality suggested by proponents of the spatial-visual skills to mathematical reasoning ability link is still an open question. It is quite possible that the unique link of mathematical reasoning to spatial-visual skills is operative only at the higher levels of mathematical reasoning. Perhaps at the levels of mathematical reasoning encountered by most high-school students, the link between verbal abilities and mathematical reasoning is just as powerful. If this were the case, then we would expect spatial skills to become an increasingly important skill for mathematical achievement and verbal skills to become less important as students move into more advanced mathematics courses. The findings of Fennema and Sherman (1977) provide support for this suggestion, but more work is needed.

Several investigators have addressed the question of whether the relation between spatial skills and mathematical achievement varies across the sexes. No consistent findings have emerged: Sherman (1980*a*) found the relation to be stronger among girls. In contrast, Hyde, Geiringer, and Yen (1975), Steel and Wise (1979), and Connor and Serbin (1980) have all found the relation to be stronger among boys. (In fact, Steel and Wise 1979 found attitudinal factors to be a stronger predictor of mathematical achievement than spatial skills for girls; and Connor and Serbin (1980) found verbal abilities to be a stronger predictor than spatial skills for girls). Finally, Fennema and Sherman (1977, 1978) and Burnett, Lane, and Dratt (1979) found no sex difference in the strength of the relation.

The issue most central to this discussion is the question of whether the sex difference in spatial ability mediates the sex difference in mathematical achievement. Basically one approach has been used to answer this question, namely, an evaluation of the effects of statistically partialling out spatial skill differences on the pattern of mathematical achievement scores. By and large, sex differences in mathematical achievement scores are significantly reduced or eliminated when spatial skills are partialled out (Burnett, Lane, and Dratt 1979; Fennema and Sherman 1977; and Hyde, Geiringer, and Yen 1975). However, as Burnett, Lane, and Dratt (1979) point out, one can not conclude from these results that the spatial-skill differential is causing the mathematical-achievement differential. One can conclude only that the data are consistent with that hypothesis. The findings reported earlier on the relations of verbal skills to girls' mathematical achievement scores make this caveat even more critical. Additionally, Fennema and Sherman (1977) found that the sex differences in mathematical achievement can also be statistically eliminated by partialling out either the number of courses taken or a set of attitudes toward math that are sex differentiated.

What can we conclude? There are sex differences in both mathematical achievement and spatial skills among eleventh and twelfth graders. These differences seem to persist into adulthood in those limited populations studied (primarily college students). Whether or not the sex difference in spatial skills is contributing to the sex difference in mathematical achievement is still an open question. The pattern of findings to date is consistent with that hypothesis, but a causal relationship has yet to be established.

Further, whether or not the sex differences in either mathematical ability or spatial skills are contributing to the sex differences in participation in or attitudes toward mathematics is an even more open issue at present. Attitude differences emerge at a younger age than do the sex differences in either achievement or spatial skills. While it is possible that the girls are already sensing that they are less "able" in mathematics in spite of the fact that they are doing just as well as the boys, it seems more likely that the drop in girls' attitudes is a consequence primarily of social factors.

In contrast to the drop in girls' attitudes toward math, the drop in girls' participation rate is more likely to reflect, to some degree, whatever sex differences in mathematical aptitude are ultimately uncovered. In the few studies that have attempted to predict participation rates in high-school students, spatial visualization sometimes emerges as a significant predictor, though not always, for both girls and boys (Sherman, in press, spatial skills predicted for girls only; Wise, Steel, and MacDonald 1979, spatial skill predicted for boys only). Participation is also predicted by scores on vocabulary tests (Sherman, in press), by past math achievements (Fennema 1981; Parsons et al. 1981; Armstrong 1980; Wise, Steel, and MacDonald 1979; Dunteman, Wisenbaker, and Taylor 1979), by interest in mathematics and career plans (see Fennema 1981; Parsons et al. 1981; Wise, Steel, and MacDonald 1979) and by a variety of attitudinal and social factors that will be reviewed in the next section. Additionally, spatial visualization skills can be trained. Thus, the magnitude of the contribution of biological factors, the inevitability of their effects, and the exact nature of these effects are still to be determined.

Biological Theories for the Sex Difference in
Spatial Skills

Setting aside the issue of whether or not sex differences in spatial skills underlie sex differences in quantitative skills or interests, the pattern of sex differences in spatial skills is certainly consistent enough to suggest a biological mediator. Several have been proposed, of which three have received the most attention: (a) an X-linked recessive gene, (b) brain lateralization, and (c) hormonal effects on cognition.

Recessive Gene Hypotheses. Geneticists have suggested that there might be a recessive gene on the X chromosome that has a positive influence on spatial perception. Since males have only one X chromosome, if they receive the recessive gene it would express itself as exceptional spatial perception. In contrast, since girls have two X chromosomes, they would need two recessive genes in order to have exceptional spatial perception. Consequently, since the likelihood of getting two recessive genes is less than the likelihood of getting one recessive gene, males on the average should have a greater chance than females of developing exceptional spatial perception. Thus, if it could be demonstrated that spatial perception is influenced by a recessive gene on the X chromosome, then one would have uncovered a biological basis for the average advantage males have on tasks involving spatial perception (Stafford 1961).

To test this hypothesis, the intrafamilial correlations of performance on spatial tasks have been examined. Since sons receive their X chromosome from their mothers and not their fathers, correlations of performance between sons and mothers should be much higher than the correlations between sons and fathers. Data from early studies confirmed this prediction (Bock and Kolakowski 1973; Corah 1965; Hartlage 1970). However, two recent studies with very large samples have found no evidence for the X-linked recessive gene hypothesis (DeFries et al. 1976; Williams 1975). Thus, present evidence seems to discount the hypothesis of an X-linked recessive trait of high spatial ability.

Brain Lateralization Hypotheses. Another biological explanation for sex dimorphic spatial abilities is differential brain lateralization. The human brain is divided into two hemispheres: the right and the left. Recent studies on split-brain subjects (individuals whose hemispheres have been separated) suggest that each hemisphere of the brain specializes in certain abilities, the left hemisphere specializing in verbal abilities and the right hemisphere specializing in spatial perception (Levy-Agresti and Sperry 1968; Sperry and Levy 1970). At some point in development, lateralization (the specialization in the functioning of each hemisphere) begins, and one hemisphere, usually the left, becomes dominant in its control of an individual's behavior. It has been argued that the timing of this lateralization may affect the development of both spatial and verbal skills. Since the most consistent sex differences in cognitive functioning are found on tasks involving either spatial or verbal skills, it has been suggested that differential timing of lateralization might underlie, to some extent, these sex differences (Harris 1978). The reasoning goes something like this: Males perform spatial tasks better than females, and females perform verbal tasks better than males. There is lateralization of the brain in relation to these two skills. Lateralization may begin earlier in females than males. Perhaps delayed lateralization gives males an advantage on spatial skills, while early lateralization gives females an advantage on verbal skills.

The sex difference in the timing of lateralization does receive fairly consistent support. Several developmental studies suggest that lateralization begins earlier in girls (Kimura 1967; Knox and Kimura 1970) although this claim is still quite controversial (see Bryden 1979; Maccoby and Jacklin 1974).

The findings with regard to more complete lateralization in postpubertal males are more consistent (see McGlone 1980; Bryden 1979). When sex differences emerge, postpubertal males exhibit a pattern of responses congruent with the hypothesis that their brains are more lateralized for both verbal and spatial-visual information processing. Bryden (1979) has offered three possible explanations for this difference. "First, there may be a real biological difference in cerebral organization between males and females, so that cognitive and perceptual functions are more likely to be bilaterally represented in females than in males" (p. 138). Based on her review of the clinical studies of brain-damaged patients, McGlone (1980) concluded that this hypothesis now has sufficient support to be taken seriously. While the majority of the commentors on her review agreed at least in part with this conclusion, several did not (for example, Denenberg 1980; Fairweather 1980; Kinsbourne 1980; Sherman 1980b). Thus, whether males' brains are structurally lateralized to a greater extent than females is still an open question.

As a second possible explanation, Bryden (1979) suggested that the observed differences might arise from the test procedures employed. Females may use different strategies when performing the tasks used to measure cerebral lateralization. Sherman (1971), Harris (1978), and Rudel, Denckla and Spaltar (1974) have all made a similar suggestion, namely, that the sex difference in spatial skills may result from females relying on a verbal rather than a spatial mode in solving spatial-visualization tasks. Since verbal strategies are less efficient than spatial strategies for these tasks, females will perform more poorly than males, especially in comparison to their relative performances on verbal tasks. Both Bryden (1979) and Rudel, Denckla, and Spaltar (1974) have found some empirical support for this suggestion. But the extent to which this explanation accounts for the apparent sex differences in brain lateralization is still unknown.

Bryden's (1979) third suggestion is the most plausible and the most complex: specifically, Bryden suggests that sex differences in apparent brain lateralization reflect the interactive effect of strategy differences and cerebral organization. "That is, perhaps females pursue different strategies . . . because their cerebral organization is different" (p. 138) and conversely, perhaps adopting different strategies solidifies or augments the magnitude of cerebral organizational differences. Harris (1978) has made a similar suggestion and there is some rudimentary empirical support for this interactionist position (see Bryden 1979). But the exact nature of differences in cerebral organization and the nature of the interactions between experience, available cognitive strategies, and cerebral organization are as yet unknown. The relations will be quite complex because the interactionist position is really suggesting that not only is the interaction of specific experiences with specific cognitive skills important but so also is the interactions among various cognitive skills and cerebral organization in determining the cognitive strategy which an individual will use for any given problem situation.

Hormonal Hypotheses. Several researchers have suggested the possibility that hormones may be implicated in the sex dimorphism of cognitive skills, especially spatial skills. The hypotheses have taken two basic forms. The first set of hypotheses focuses on the impact of hormones on the brain during the prenatal period; sex-differentiated exposure to prenatal testosterone, it has been argued, might account for later sex dimorphism in cognitive functioning. Early studies of adrenogenital syndrome girls (girls who have been exposed to unusually high levels of an androgen due to genetic problems) and of the effects of exposure of the mother to exogenous, androgen-like compounds during pregnancy have yielded some support for this suggestion. These studies indicated that the females exposed to such compounds had slightly higher IQs than normal (see Ehrhardt and Meyer-Bahlburg 1981). However, recent studies with appropriate controls have failed to find any relation between prenatal exposure to abnormally high levels of androgens and later cognitive functioning.

Two other clinical syndromes have yielded an interesting pattern of results related to the hypothesized relation between prenatal hormones and subsequent cognitive functioning. Both androgen-insensitive (XY) females (individuals who are anatomical females because their bodies are insensitive to androgen) and Turner syndrome females (individuals who are anatomical females because they lack a second sex chromosome) do relatively poorly on spatial tasks (see Reinisch, Gandelman, and Spiegel 1979). These groups were exposed prenatally either to very low levels of androgens or to no functional androgens. The pattern of their performance scores is consistent with the hypothesis that exposure to prenatal androgens facilitates later performance on spatial tasks. The results, however, are also compatible with other explanations. The evidence from the Turner syndrome females is especially difficult to interpret since the level of their exposure to prenatal hormone is confounded with the absence of a second sex chromosome.

Reinisch, Gandelman, and Spiegel (1979) have offered an alternative suggestion regarding the effects of prenatal hormones on later cognitive function that could explain both the general patterns of sex dimorphism in spatial skills and the specific patterns exhibited by individuals with either of these two syndromes. Since exposure to prenatal androgens is related positively to activity level in childhood (this will be discussed more extensively later), Reinisch and his co-workers argue that activity-level differences resulting from variation in the level of prenatal androgens could produce differences in experience that, in turn, could account for the sex dimorphism found on tests of spatial skills. Consistent with this point of view, Connor, Serbin, and Schackman (1977) have suggested a relation between preschoolers' preference for large-muscle, exploratory play and their performance on tests of spatial skills. The success of training studies in increasing females' performance on tests of spatial skills also provides support for the importance of experience in developing spatial skills (see Connor, Schackman, and Serbin 1978; Goldstein and Chance 1965). However, whether sex differences in activity level per se are sufficient to produce sex-dimorphic performance on tests of spatial skill is still unknown. What is most interesting about Reinisch, Gandelman, and Spiegel's (1979) hypothesis is that it points out quite clearly that one must entertain a very broad view of the possible interactions between biological processes and experience in shaping behavior.

The second set of hormonal hypotheses focuses on the possible direct effects of postpubertal hormone level on cognitive functioning. Broverman and his colleagues (1980) have carried out an extensive investigation of the relation between sex hormones (primarily androgens) and cognitive performance. Most of their work has focused on males. Within that body of research it is now fairly clear that relatively high levels of testosterone impede the performance of males on the kinds of tasks commonly used to measure spatial visualization (see Broverman, Klaiber, and Vogel, 1980). In a study using the degree of androgenization of one's body as an indicator of testosterone levels, Petersen (1979) found comparable results for males; the males who had the least androgenized body type in terms of muscle mass, body shape, penis size, and pubic hair distribution did the best on tests of spatial ability. However, for females, the results ran counter to the effects found in males; the females who had the most androgenized body type did the best on the tests of spatial skills. Petersen (1979) concluded that "androgynous males and females tend to excel at spatial ability . . . whereas individuals who are more sex-stereotypic in appearance tend to do poorer at spatial ability'' (p. 204).

Thus, a simple hormonal explanation of sex dimorphism on spatial tasks seems unlikely. It is not the case that low levels of androgens are always associated with spatial skills. If that were true then females would have higher spatial skills than males. Nor is it the case that patterns of relations will be the same in males and females. As Petersen (1979) suggested, the effects of hormones on cognitive functioning are quite complex and probably interact with several mechanisms such as maturational rate, body type, brain lateralization, and the social channeling of experiences resulting from early or late maturation and body type. For example, Waber (1979) has demonstrated that late maturation facilitates spatial skills. Late maturers also tend to have greater brain lateralization and more androgynous body types (that is, tall, slender, and athletic). While hormones may be implicated in each of these processes, they may or may not be a cause of sex dimorphism in spatial skills. Given the similarity of these relations, it is difficult to separate out what is actually causing what and how these variables might be interacting with each other or with other causally critical variables.

In conclusion, a variety of biological processes are being investigated in the attempt to explain sex differences in cognitive functioning. The research is still in its infant stages; no definitive answers are now available. The following conclusion, however, seems clear at present: (1) the factors shaping sex differences in cognitive functioning will be complex and highly interactive; (2) different factors may shape the performance of males and females; (3) different factors may shape performance at different ages; and (4) experience in interaction with biology will play a major causal role.

Sex Differences in Parenting

The consistency of both cross-cultural and cross-species behavior suggests that a biological component may be associated with the sex-dimorphic pattern of parenting. Parenting in both primates and humans is generally the female's job. But whether this role assignment is biologically based is an extremely difficult question. In considering cross-cultural patterns, for example, it must be noted that there is considerable overlap in both the socialization patterns and the ecological realities to which cultures must adapt. For example, while women are the child-raisers in most cultures, at the same time most cultures actively socialize girls into this role (Barry, Bacon, and Child 1957). Additionally, in many cultures mothers are the primary protein food source for infants and thus, by necessity, must be assigned the role of nursing the infants. Since women are needed to nurse the infants they bear, and since contraceptives are often not readily available, women have to spend most of their adult years around the children. It makes practical sense, then, to assign them the role of raising the children. Thus, it is difficult to know whether this division of roles was selected according to evolutionary factors or whether it reflects a common solution to a common survival problem (Archer 1976).

Such overlaps in socialization patterns, ecological demands, and cultural universals make evaluation of the relative importance of socialization and biology very difficult. Socialization could be producing the differences, it could be exaggerating a small biologically based difference, or it could be mirroring a powerful biologically determined behavioral system (Archer 1976; Goldberg 1973; Maccoby and Jacklin 1974; Reiter 1975, 1976; Rosaldo and Lamphere 1974). Because these distinctions between relative weighting are crucial in our conceptualization of sex-role malleability, they have important implications for social change. Unfortunately, for most behaviors, including parenting, it is not possible as yet to decide this issue, and much of what is being debated today is primarily speculative. Having women do the major parenting is adaptive from both a cultural and an evolutionary perspective. Thus, it is plausible that both biosocial and socialization forces are pushing women to fulfill this role. The role of socialization has been demonstrated time and time again; the role of biosocial forces is much harder to assess.

In one of the most persuasive discussions of this issue, Rossi (1977) argued that evolutionary forces have selected for heightened maternal investment in children, greater propensity for acquiring parenting skills in females, and reciprocal, physiologically based bonding systems in both infants and mothers. The evidence she cited speaks most directly to the last of these three, namely, the physiologically based bonding system. There are physiological events associated with early attachment. For example, an infant's cry stimulates the mother's secretion of oxytocin, which, in turn, prepares her breasts for nursing. The hormone oxytocin is also involved in sexual responsiveness. Thus, there is a link between the sexual response and the lactation system such that nursing can produce enjoyable sexual sensations. On a more behavioral level, Rossi cited several studies indicating that there may be a biosocial component in early attachment. For example, mothers regularly exhibit a fixed sequence of behaviors when they first explore their new infants. In addition, Klaus and Kennell (1976) have argued that mothers are in a "sensitive period" just after birth such that exposure to their infants then solidifies the mother-infant attachment bond. Supposedly a solid bond facilitates later interaction while disruption of the early contact between mothers and infants may have long-range effects on mother-child attachments evidenced by such behaviors as child abuse and neglect. While early studies provided some support for Klaus and Kennell's suggestion, more recent studies have failed to demonstrate any long-term advantage of early contact that cannot be accounted for by the general excitement associated with the birthing process and the positive social and emotional milieu associated with those early days of parenting (see Lamb and Goldberg, 1980; Parke and Sawin 1977). These processes are potentially available to fathers as well as mothers.

Thus, while it does seem that biosocial forces may be involved in mother-child bonding, the nature of these forces and the extent to which they operate differentially in men and women are unknown. While evidence from the animal literature suggests that hormonal changes associated with pregnancy and parturition might be involved (see Lamb 1975), generalizations across species are problematic. This is especially true for parenting since there is little similarity between human parenting behaviors and the parenting behaviors of rodents. Until more work is done with higher primates and humans, the role of hormonal shifts in "priming" parenting behavior in human mothers is still unknown. And it seems likely that whatever hormonal effects may emerge, their impact on mother-child bonding has undoubtedly been heightened by a heavy overlay of socialization pressures.

We know very little about bonding in fathers. Because child-rearing is assumed to be the domain of women, bonding processes have rarely been studied in men. Further, in many cultures, fathers are systematically excluded from the birth process and from early contact with infants. If bonding is affected by early contact (still a debatable hypothesis), then cultures effectively block the natural attachment between fathers and their infants. Some evidence does, in fact, suggest that early contact between father and infant affects subsequent measures of attachment in the predicted direction (see Lamb and Goldberg 1980). Thus, it is not yet possible to assess the *extent* to which biosocial forces foster parent-child bonding as opposed to mother-child bonding. An examination of cross-cultural and cross-species fathering can at least provide some insights into the potential for and range of expression of father-child attachment. By showing the range of potential father-child involvement, we can at least speculate on the possible malleability of parenting role assignments.

The degree of paternal involvement in the parenting of higher primates is quite variable. In some species, for example, rhesus monkeys, baboons, and chimpanzees, males play little if any direct role in parenting; in other species, for example many New World monkeys, particularly marmosets, males play a very active parenting role (Redican 1976). Further, there have been instances in which the males of a low-paternal species exhibited a high degree of involvement when the situation warranted these behaviors.

The range of parenting behaviors is also quite broad. According to Redican (1976), males exhibit, albeit with lower frequency, the full range of parental behaviors commonly exhibited by females; for example, they "premasticate food for infants"; "carry, sleep with, groom, play with" and teach the young; and provide refuge "during periods of high emotional arousal" (p. 378). In addition, they exhibit the behaviors commonly associated with the male protector role. Thus it seems that paternal involvement is clearly within the repertoire of behaviors available to high primate species.

Are there factors that influence the extent of paternal involvement of primates? Redican (1976) suggests that the following factors influence the involvement of primate fathers in raising offspring: (a) monogamous social organization; (b) availability of stable food supply; (c) low levels of competition and hostility between different social groupings of the same species; and (d) relaxed, permissive maternal-infant interactions. These structural characteristics suggest that paternal involvement is high when paternity is readily identifiable, when males are not needed for the warrior-hunter role, and when females tolerate and encourage male parenting.

Can one generalize the findings concerning primate parenting to humans? With results similar to Redican, West and Konner (1976) conclude that plasticity in the extent and form of paternal behavior is also characteristic of human males. Like primates, human males are universally less involved in parenting than females, but they too exhibit a wide range of parenting behaviors when necessary. West and Konner (1976) suggest the following structural arrangements as facilitative of human paternal involvements: (a) monogamy; (b) nuclear family units; (c) low levels of local warfare; (d) maternal employment; and (e) a gathering and/or agricultural economy. As is the case with the subhuman primates, then, human paternal involvement is increased by easily identifiable paternity, low demand for the warrior-hunter role, and high opportunity and need for father-child interaction. Men take care of their children if they are sure they are the fathers, if they are not needed as warriors and hunters, if the mothers contribute to family resources, and if their parenting is both necessary and encouraged.

But, even when all of these conditions are present, men still play a less active role than women in child-rearing. Is this difference biologically based? We do not know. On the one hand, there is some evidence that testosterone lowers maternal behavior in lower animals (West and Konner 1976). Further, studies of adrenogenital syndrome females have repeatedly found that girls who have been exposed to unusually high levels of prenatal androgens are less interested in doll play than are the "normal" controls (see Ehrhardt and Meyer-Bahlburg, 1979, 1981). Whether these effects are related to adult parenting behavior, and whether they are a direct consequence of hormonal exposure or an indirect consequence of other variables such as activity level, are still unanswered questions.

On the other hand, neither socialization pressures nor birthing practices encourage paternal involvement. For example, Ember (1973) found that helping to take care of younger children increases nurturing and socially responsible behaviors in boys. Whether these boys will exhibit more paternal behaviors as adults is yet to be seen. But if they do, then early experience with caring for younger children may be another of those precursors of "maternal" caring that is generally denied to males.

Similarly, several recent laboratory studies of parenting behavior have uncovered a pattern of results that run counter to the theme of biologically based sex differences in parenting. For example, Frodi and Lamb (1978) compared natural, unlearned psychophysiologically based responses with overt, learned behavioral responses. While the males and females in this study did differ in their overt behavioral response to infants, they did not differ in their psychophysiologically based responses. These results suggest that while the physiological responses associated with responding to infants are present equally in both males and females, males and females differ in their behavioral responses to these physiological cues. Females appear more likely to respond with parenting-like behaviors while males are more likely to ignore or withdraw from the infants. Studies with parents demonstrate even more clearly that both fathers and mothers have the capacity to parent. Fathers have been found to be as capable as mothers in performing child-care activities, and as sensitive as mothers in their responses to infant signals (Parke and Sawin 1977). Further, mothers and fathers display identical physiological responses to both infant cries and infant smiles even though the mothers report more extreme emotions (Frodi et al. 1978a, b).

What could account for this pattern of results? A recent series of studies by Feldman and her colleagues (Feldman and Nash 1978; Feldman, Nash, and Cutrona 1977; Abraham, Feldman, and Nash 1978) and by Frodi and Lamb (1978) demonstrate that the expression of parenting-like behaviors is under strong social control. For example, behavioral responsivity varies with life stage such that a sex difference is present at puberty and among very new parents but not among college students, young adults, childless couples, and cohabiting adults. Further, men and women report more attraction to babies in same-sex groups than in mixed-sex groups, and the likelihood of public disclosure increases women's expressed attraction to babies and decreases men's (Berman et al. 1975). These findings suggest that the overt expression of parenting behaviors are linked to sex-role socialization and to the salience of one's need to appear appropriately sex-typed. Whether the expression of parenting behaviors is also under biological control is an open question.

An alternative explanation for the lower participation of fathers in parenting is suggested by the dialectical perspective. Perhaps sex dimorphism in parenting is not a consequence of differences in response to infants alone but instead is a consequence of the broad array of tasks that involve men and women. In addition to the cross-cultural consistency in the assignment of parenting to mothers, the warrior-defender-provider role is usually assigned to males. If you recall, fathers are more involved in child care when there is relatively little need for their involvement in the warrior-defender role and when the provider role is shared more equally with mothers. This pattern suggests that the factors influencing sex dimorphism in those behaviors commonly associated with males also influence men's involvement in the parenting role.

In line with this reasoning, it is also possible that men and women are equally invested in their children but express their investment in different ways. Most studies of parental investment have defined investment in terms of typical maternal behaviors. Few studies have attempted to assess investment in terms of male values or male behaviors. Can we conclude that males are less invested in their children if they are not actively involved in day-to-day child-care? Men may express their investment in their children through their provider-protector role rather than through a nurturing parent role. Children serve many different needs for adults; some needs are more typical of women while others are more typical of men (Hoffman and Hoffman 1973). Assessing the differential subjective importance of these various values will be an extremely difficult task.

In conclusion, it appears that adaptability in parenting styles for both males and females is as much a part of our biosocial heritage as is heightened maternal investment in children. In addition, it is clear that investment can be expressed in a variety of ways. More research is needed on the whole range of relevant behaviors, and on a wider array of possible biological inputs and mechanisms of interaction between biology and experience.

Aggression

For the most part, reviewers and researchers have both concluded that adult males (both humans and primates) generally exhibit more intraspecies physical aggression (see Archer 1976; Frodi, Macaulay and Thome 1977; Reinisch 1981; Rosenblatt and Cunningham 1976). They are far more likely than females to be involved in combat and in various other forms of antisocial aggression. For example, as noted earlier, the warrior-provider role is virtually always assigned to the male. Similarly, men are much more likely to be involved in violent crimes than women. In fact, while in recent years the general crime rate has been rising more rapidly among women than among men, the incidence of violent crime is still rising more rapidly among men (Newsweek 1975). Measures reflecting attitudes toward aggression reveal a similar pattern; Gallup polls have repeatedly indicated that women in the United States are much less likely than men to endorse either military involvement of any kind, or capital punishment (see Frieze et al. 1978). Finally, younger males also typically exhibit more rough-and-tumble play and more physical aggression than females (see Maccoby and Jacklin 1974; Whiting and Edwards 1973).

As consistent as these global patterns are, however, males are not always more aggressive than females. Recent reviews of the available literature suggest notable exceptions that are critical to any understanding of the determinants of aggression. According to Hoyenga and Hoyenga (1979), both the size of the animal and the reproductive state of the female must be taken into account. Sex dimorphism in aggression is much less marked among species in which males and females are of approximately equal size. Additionally, lactating females have been known to be quite aggressive in defense of their young (Floody and Pfaff 1977). Frodi et al. (1977) pointed out several other exceptions. In particular, they noted that female aggression goes up in situations in which aggression is condoned and in which cues likely to elicit empathical responses are minimized. Finally, several reviewers have argued that we have not yet studied the full range of aggressive behaviors. Researchers have tended to focus on those forms of aggression that are more characteristic of males—a problem that stems from the basic difficulty in defining aggression. Consequently we do not really know whether sex differences exist in many forms of aggressive behavior. Nonetheless, at least in terms of physical aggression and open displays of hostility, the pattern of results is consistent enough to entertain biological hypotheses. It is to these hypotheses that I now turn.

Hormonal Hypotheses

Two basic modes of hormonal influence have been proposed: an inductive mode and an activation mode. Hormonal induction refers to the process by which prenatal or perinatal hormones affect the development of the brain. Critical embryological periods have been identified in most species. In subhuman primates, exposure to the appropriate gonadal hormones at that period appears to "masculinize" the brain, increase the frequency of masculine behavior patterns, and sensitize the brain to postpubertal exposures of various gonadal hormones. Hormonal activation refers to the process by which exposure to different gonadal hormones affects an animal's ongoing behavior. With regard to gonadal hormones, activation usually occurs postpubertally.

There is ample evidence in subhuman mammals that prenatal and perinatal exposure to androgens influences aggressive behavior patterns. Both genetic male and female animals exposed to androgens at the critical period exhibit high levels of rough-and-tumble play as juveniles (Young, Goy and

Phoenix 1964) and other forms of aggressive display as adults (see Money and Ehrhardt 1972; Reinisch 1981). In addition, sensitivity to androgens postpubertally is also affected. Animals exposed to androgens pre- or perinatally typically respond with increased aggression and activity level when given androgens postpubertally (see McEwen 1981; Money and Ehrhardt 1972; Reinisch 1981). Interestingly, in some species exposure to estrogens produces comparable results (Bronson and Desjardins 1968).

Whether prenatal exposure to androgens has a comparable effect in humans is a difficult question to answer, primarily because it is unethical to run comparable studies with humans. Instead, scientists have had to rely on "natural" experiments—experiments in which the prenatal hormonal environment has been varied for some "natural" reason. Ehrhardt, Money, Meyer-Bahlburg, and Reinisch have used these naturally occurring deviations from the normal pattern of sexual differentiation to assess the possible impact of prenatal hormones on human sex-role dimorphism.

A word of caution is necessary before beginning this review. Some of these studies are based on a small number of clinical cases in which subjects differed from a "normal" sample in several important ways, for example, prenatal hormonal history, appearance of genitalia at birth, and membership in a clinical population. Given the uniqueness of these individuals, generalizations must be made with extreme caution. In addition, the causal origins of the behavioral patterns in these samples are unclear. The patterns could have resulted from their exposure to the prenatal hormones, from their familiarity with the clinical setting, from their awareness of their own uniqueness, from the reactions of others who know about their unique status, or from some interaction of two or more of these. The more recent work of Reinisch, and of Ehrhardt and Meyer-Bahlburg, has avoided some of these problems by studying populations without abnormal genitalia at birth.

The classic work in this area (that of Money and Ehrhardt) involved children suffering from the adrenogenital syndrome. These children had been exposed to unusually high levels of androgen or androgen-like components prenatally either because their own adrenal systems were malfunctioning or because their mothers were given synthetic progestins to prevent miscarriage. The girls in these studies typically had "masculinized" genitalia at birth and were involved in clinical treatment of varying degrees. Both adrenogenital-syndrome boys and girls reported higher than "normal" involvement in active sports.

The more recent work of Ehrhardt and Meyer-Bahlburg (1979) and Reinisch (1981) has focused on children whose mothers were exposed to a synthetic progestin that has androgenizing properties in the fetus. Children in these studies did not have abnormal genitalia and were not undergoing clinical treatment. By and large results similar to those reported above have emerged. These children exhibit higher levels of activity usually in the form of participation in active sports. One study (Zussman, Zussman and Dalton 1977) found that boys of mothers given a natural progesterone during pregnancy were more aggressive in school and more likely to get into trouble as a result than "normal" boys. Finally, Reinisch (1981) found a higher level of self-reported hostility and preference for physically aggressive solutions to problems than their sib-controls among both boys and girls whose mothers were given synthetic progestins.

These studies suggest a link between prenatal exposure to hormones and subsequent behaviors. But while the androgenized children did evidence higher activity levels, they did not in general exhibit more physically aggressive behavior. This failure to find a significant increase in aggression suggests that socialization dictates the expression, if not the emergence, of any potential for greater aggressive behavior that might be created by prenatal androgens. Perhaps instead of physical aggressiveness per se, prenatal androgens predispose the developing organism to a higher level of physical activity, the exact manifestation of which is dependent on socialization (Frieze et al. 1978). Alternatively, prenatal androgens may create a potential for aggressiveness that requires postnatal androgens for its expression. Consequently, since these females are being treated and therefore are not being exposed to postnatal androgens, it would not be expected that they would exhibit the typically high level of aggression displayed by males (Frieze et al. 1978).

Work demonstrating the activation effects of gonadal hormones has also revealed a consistent pattern among subhuman mammals. Aggression in normal males is increased by exposure to androgens; aggression in females does not appear to be as much under the control of androgens as male aggression unless the female is exposed to high dosages over a prolonged period of time; and finally, female aggression does seem to be somewhat under the control of hormonal variations associated with lactation (see Hoyenga and Hoyenga 1979 for a full review). Even these effects, however, are subject to social influences. The relation between aggression and androgens is lower among the more social species, among more mature animals, and after certain kinds of experience that are typically related to an animal's position in the dominance hierarchy. Further, in some species the effect of androgens on aggression depends on the levels of other circulating hormones; in particular, the female hormones. Finally, in subhuman primates it is clear that experience can have as much of an effect on androgen levels as androgen levels can have on behavior. In particular, both stress and dominance position affect testosterone levels (Macrides, Bartke, and Dalterio 1975; Rose, Holaday, and Bernstein 1979).

A similar pattern of mixed findings has emerged with humans. While Persky, Smith, and Basu (1971) found a positive relation between levels of circulating androgens and self-reported projective measures of hostility and aggression, these results were more characteristic of the younger men in their sample. Subsequent studies have failed to replicate their result (see Ehrenkranz, Bless, and Sheard 1974; Kreuz and Rose 1972; Persky et al. 1977). Similarly, while some studies report that injections of testosterone lead to an increased level of aggression, activity, and a sense of well-being, other studies indicate that injections of antiandrogens do not significantly reduce aggressive behaviors in criminal and mentally ill populations unless massive dosages are used (see Rubin, Reinisch, and Hasbett 1981). Finally, while criminals imprisoned for violent crimes do tend to have higher levels of circulating testosterone than do criminals imprisoned for nonviolent crimes, the levels of testosterone do not correlate with ongoing levels of aggressive behavior at the time the measures are taken (Kreuz and Rose 1972).

Males and females differ on one other set of characteristics that is also related to androgens; namely, body characteristics. Males have body characteristics that may suit them better for physical aggression. They are bigger, have more muscle mass, have higher metabolism rates, and have a higher proportion of red blood corpuscles (Schienfeld 1958). Given the findings in the animal literature that larger animals are typically the more aggressive, sex differences in size could certainly contribute to the sex differences in aggression.

Reviewing this body of literature, Hoyenga and Hoyenga (1979) concluded that testosterones, if they have any causal impact, appear to have that impact during adolescence. The relation between aggression and testosterone is reduced during later adulthood, perhaps due to the effects of experience and socialization. Whether this is indeed true, however, is still unknown. As Petersen (1980) concludes, the relation between aggression and androgens in humans is unclear at present, but is likely to exist at some level and to be highly subject to learning and to environmental influences.

Other Hypotheses

Two other hypotheses have been offered for sex difference in aggression: (a) inhibition of aggression in women by other responses such as guilt and empathy, and (b) infant-parent interactions that dialectically evolve into a pattern likely to encourage greater aggression in boys. Each of these are discussed briefly below.

Several reviewers have suggested that females are less aggressive because they are more of something else. The most common competing responses are empathy and guilt. As Frodi, Macauley, and Thome (1977) point out, there is enough evidence from laboratory studies of aggression to support the conclusion that levels of aggression are affected by empathic responses and by guilt, and that females appear to be more subject to these effects than are men. Additionally, there is some evidence to suggest that women's physiological responses to their own aggressive acts are different from those of men. Women do not display the physiological signs of relief as readily as do men following a counteraggressive act (see Frodi, Macauley, and Thome 1977 for full review). Consequently, it is possible that aggression in women is inhibited both by empathic responses prior to an aggressive act and by the guilt feelings that follow such an act. Together these two mechanisms would certainly reduce the likelihood of aggressive behaviors in women. But how these sex differences are acquired is not known, and the possible biological mediators have yet to be studied.

Another hypothesis, first proposed by Bell (1968), is based on the assumption that infant characteristics like irritability and activity level influence the response of parents. Parental responses, in turn, shape further infant development, which, in turn, shapes parent-infant interactions. Through their impact on this cyclical process of developing interactive patterns, then, individual differences in early infant characteristics can "create" individual differences in major behavioral patterns later in life. One specific example used by Bell to illustrate this process is the sex difference in aggression. He argued that, if boys are more irritable at birth, then one could expect a more negative infant-parent interaction to evolve and, consequently, boys would become more aggressive than girls. The process he described is outlined in table 8–1.

Table 8–1
Possible Interactions between Baby's Characteristics and Parental Responsiveness

Baby A (More likely to be a girl)	Baby B (More likely to be a boy)
Baby's characteristics	
Physically mature	Physically immature
Sleeps a lot	Cries a lot
Vocalizes to faces	Active and therefore gets into trouble
Smiles at faces	
Parent Response	
Affectionate	Irritable
Responsive when child does cry	Not necessarily reponsive to child's frequent cries
Talks to child	Uses physical restraints and punishments
Child Response	
Affiliative—likes people, expects people to satisfy needs	Aggressive
Early vocalization	Expects to get needs satisfied through own efforts

Source: I.H. Frieze, J.E. Parsons, P. Johnson, D.N. Ruble, and G. Zellerman, *Women and Sex Roles,* New York: Norton and Company, 1978, p. 78. Reprinted with permission.

Some evidence suggests that male infants are fussier or more irritable than female infants, for example Moss (1967). The results depend, however, on a variety of other factors, such as specific age at testing, prenatal and delivery complications (Parmalee and Stern 1972) and birth order. Nevertheless, when differences are found, male infants are generally the more irritable. And, whatever the cause of this early difference in irritability, the crux of Bell's hypothesis is that these sex-related differences in neonatal temperament could set in motion a social interaction pattern that would result in boys being more negative, resistive, and aggressive than girls. Circular interactive processes between the parent and child thus could turn the irritable baby into the aggressive child (Bell 1968).

Conclusion

What then can we say about sex differences in aggression? Aggression is an excellent example of the dialectical model of the interaction of biology and experience in shaping behavior. We know that biological events during the prenatal period can affect expression of aggressive behavior later in life. We also know that experience has a major impact on not only the expression of aggression but also on the biological system itself. For example, it is clear that stress and dominance placement in both humans and subhuman mammals affect levels of hormone production in males. Thus, the link between behavior and hormones goes in both directions. Further, we know that experience can alter the degree of the relation between hormones and behavior. For example, castration after puberty has a much less dramatic effect on aggressive behavior than does castration prior to puberty (see Hoyenga and Hoyenga 1979). We also know that the expression of aggression is multiply determined, that it is under the influence of a variety of factors including socialization pressures, the intercession of incompatible emotions such as guilt and empathy, and social norms, and that it can take a variety of forms ranging from murder to weekend college football. Finally, we know that the origin of the sex difference is multiply determined and is influenced not only by a variety of social events but also by a variety of biologically initiated processes including prenatal hormones and maturational rates. Like the two clusters of behaviors discussed earlier, aggression is a complex behavior that is shaped by a multitude of processes that wax and wane across each individual's life span.

Summary

This chapter began with two major goals: (a) the presentation of a dialectical model for the interaction of biology and experience in shaping behavior; and (b) the presentation of three examples of these processes that relate directly to our understanding of the origins of sex-dimorphic behavior. Each of the examples (cognitive functioning, parenting, and aggression) were illustrative of the complex interaction between a wide range of forces in shaping human behavior. In addition, each example demonstrated the range of forces that might be responsible for channeling behavior along sex-dimorphic paths. It should be noted before closing that the sex differences in each instance are small, that variability in the expression of any of these behaviors is the norm, and that for the most part variability within each sex is as great, if not greater, than the average differences between the sexes.

This brings us back to the issue of malleability of behavior. It is on this issue that I wish to close my discussion. The extent to which biology sets in motion the dialectical processes that, in the end, produce sex-dimorphic behavior patterns is not yet known. But to the extent that biology is an important factor, it is still not the case that sex differences are inevitable. Biology may make it easier for one sex or the other to acquire certain behaviors, or may increase the likelihood that a given stimulus will elicit a particular response in one sex or the other. Nonetheless, the range of individual differences on all these behaviors, and the degree to which social forces also shape the behaviors in the sex-dimorphic direction, suggest that both sexes can learn these behaviors. The classic work of Mead (1935) tells us one thing if nothing else—socialization can redirect development of behavior such that sex differences are either minimized or maximized.

References

Abraham, B.; Feldman, S.; and Nash, S.C. 1978. Sex role self-concept and sex role attitudes: Enduring personality characteristics or adaptations to changing life situations? *Developmental Psychology* 14:393–400.

Archer, J. 1976. Biological explanations of psychological sex differences. In B. Lloyd and J. Archer (eds.), *Exploring Sex Differences.* London: Academic Press.

Armstrong, J. 1980. Achievement and participation of women in mathematics. Final report to National Institute of Education, Washington, D.C.

Bakan, D. 1966. *The Duality of Human Existence.* Chicago: Rand McNally.

Bardin, C.W., and Catterall, J.F. 1981. Testosterone: A major determinant of extragenital sexual dimorphism. *Science* 211:1285–1293.

Barry, H., III; Bacon, M.K.; and Child, I.L. 1957. A cross-cultural survey of some sex differences in socialization. *Journal of Abnormal and Social Psychology* 55:327–333.

Bell, R.Q. 1968. A reinterpretation of the direction of effects in studies of socialization. *Psychological Review* 75(2):81–95.

Bell, R.Q.; Weller, G.M.; and Waldrop, M.F. 1971. Newborn and pre-schooler: Organization of behavior and relations between periods. *Monographs of the Society for Research in Child Development* 36(1–2, serial no. 142).

Benbow, C.P., and Stanley, J.C. 1980. Sex differences in mathematical ability: Fact or artifact? *Science* 210:1262–1264.

Berman, P.; Abplanalp, P.; Manfield, P.; and Shields, S. 1975. Sex differences in attraction to infants: When do they occur? *Sex Roles* 1:311–315.

Bock, R.D., and Kolakowski, D. 1973. Further evidence of sex-linked major-gene influence on human spatial visualizing ability. *American Journal of Human Genetics* 25:1–14.

Bronson, F.H., and Desjardins, C. 1968. Aggression in adult mice: Modification by neonatal injections of gonadal hormones. *Science* 161 (3842):705–706.

Broverman, D.M; Klaiber, E.L.; and Vogel, W. 1980. Gonadal hormones and cognitive functioning. In J.E. Parsons (ed.) *The Psychology of Sex Differences and Sex Roles.* Washington, Hemisphere Publishing Co.

Bryden, M.P. 1979. Evidence for sex-related differences in cerebral organization. In M.A. Wittig and A.L. Petersen (eds.), *Sex-Related Differences in Cognitive Functioning.* New York: Academic Press.

Burnett, S.A.; Lane, D.M.; and Dratt, L.M. 1979. Spatial visualization and sex differences in quantitative ability. *Intelligence* 3:345–354.

Condry, J., and Condry, S. 1976. Sex differences: A study of the eye of the beholder. *Child Development* 47:812–819.

Connor, J.M.; Schackman, M.; and Serbin, L. 1978. Sex-related differences in response to practice on a visual-spatial test and generalization to a related test. *Child Development* 49:24–29.

Connor, J.M., and Serbin, L.A. 1980. Mathematics, visual-spatial ability and sex-roles. Final report to National Institute of Education, Washington, D.C.

Connor, J.M., Serbin, L.A.; and Schackman, M. 1977. Sex differences in children's response to training on a visual-spatial task. *Developmental Psychology* 13(3):293–294.

Corah, N.L. 1965. Differentiation in children and their parents. *Journal of Personality* 33:300–308.

Denenberg, V.H. 1980. Some principles for interpreting laterality differences. *Behavioral and Brain Science* 3:232–233.

DeFries, J.C.; Ashton, G.C.; Johnson, R.C.; Kuse, A.R.; McClearn, G.E.; Mi, M.P.; Rashad, M.N.; Vandenberg, S.G.; and Wilson, J.R. 1976. Parent-offspring resemblance for specific cognitive abilities in two ethnic groups. *Nature* 261:131–133.

Dunteman, G.H., Wisenbaker, J., and Taylor, M.E. 1979. Race and sex differences in college science program participation. Report prepared for National Science Foundation. Research Triangle Park, N.C.

Educational Testing Service. 1979. *National College-Bound Seniors,* 1979. Princeton: College Entrance Examination Board.

Ehrenkranz, J.; Bless, E.; and Sheard, M.H. 1974. Plasma testosterone: Correlations with aggressive behavior and social dominance in man. *Psychosomatic Medicine* 36:469–479.

Ehrhardt, A.A., and Meyer-Bahlburg, H.F.L. 1979. Prenatal sex hormones and the developing brain: Effects on psychosexual differentiation and cognitive function. *Annual Review of Medicine* 30:417–430.

――――. 981. Effects of prenatal sex hormones on gender-related behaviors. *Science* 211:1312–1317.

Ember, C.R. 1973. The effects of feminine task assignment on the social behavior of boys. *Ethos* 1:424–439.

Fairweather, H. 1980. Sex differences: Still being dressed in the emperor's new clothes. *Behavioral and Brain Sciences* 3:234–235.

Feldman, S.S., and Nash, S.C. 1978. Interest in babies during young adulthood. *Child Development* 49:617–622.

Feldman, S.S.; Nash, S.C.; and Cutrona, C. 1977. The influence of age and sex on responsiveness to babies. *Developmental Psychology* 13:656–657.

Fennema, E. 1981. Attributional theory and achievement in mathematics. In S.R. Yussen (ed.), *The Development of Reflection.* New York: Academic Press.

Fennema, E., and Sherman, J. 1977. Sex-related differences in mathematics achievement, spatial visualization and affective factors. *American Educational Research Journal* 14:51–71.

――――. 1978. Sex-related differences in mathematics achievement and related factors: A further study. *Journal for Research in Mathematics Education* 9:189–203.

Floody, O.R., and Pfaff, D.W. 1977. Aggressive behavior in female hamsters: The hormonal basis for fluctuations in female aggressiveness correlated with estrone state. *Journal of Comparative and Physiological Psychology* 91:443–464.

Fox, L.H.; Brody, L.; and Tobin, D. 1980. *Women and the Mathematical Mystique.* Baltimore: Johns Hopkins University Press.

Freud, S. 1965. *New Introductory Lectures in Psychoanalysis* (J. Strachey, ed. and trans.). New York: Norton (originally published 1933).

Frieze, I.H.; Parsons, J.E.; Johnson, P.I.; Ruble, D.N.; and Fellman, G. 1978. *Women and Sex Roles: A Social Psychological Perspective.* New York: Norton.

Frodi, A.M., and Lamb, M.E. 1978. Sex differences in responsiveness to infants: A developmental study of psychophysiological and behavioral responses. *Child Development* 49:1182–1188.

Frodi, A.M.; Lamb, M.E.; Leavitt, L.A.; and Donovan, W.L. 1978*a*. Fathers' and mothers' responses to infant smiles and cries. *Infant Behavior and Development* 1:187–198.

Frodi, A.M.; Lamb, M.E.; Leavitt, L.A.; Donovan, W.L.; Neff, C.; and Sherry, D. 1978*b*. Fathers' and mothers' responses to the faces and cries of normal and premature infants. *Developmental Psychology* 14:490–498.

Frodi, A.; Macauley, J.; and Thome, P.R. 1977. Are women always less aggressive than men? A review of the experimental literature. *Psychological Bulletin* 84:634–660.

Gitelson, I.B. 1980. The relationship of parenting behavior to adolescents' sex and spatial ability. Paper presented at American Psychological Association, Montreal.

Goldberg, S. 1973. *The Inevitability of Patriarchy.* New York: Morrow.

Goldstein, A.G., and Chance, J.E. 1965. Effects of practice on sex-related differences in performance on embedded figures. *Psychonomic Science* 3:361–362.

Harris, L.J. 1978. Sex differences in spatial ability: Possible environmental, genetic, and neurological factors. In M. Kinsbourne (ed.), *Assymetrical Functions of the Brain.* Cambridge: Cambridge University Press.

Hartlage, L.C. 1970. Sex-linked inheritance of spatial ability. *Perceptual and Motor Skills* 31:610.

Herbst, L., and Petersen, A. 1980. Timing of maturation, brain lateralization and cognitive performance. Paper presented at American Psychological Association, Montreal.

Hoffman, L.W., and Hoffman, M.L. 1973. The value of children. In J.T. Fawcett (ed.), *Psychological Perspectives on Population.* New York: Basic Books.

Hoyenga, K.B., and Hoyenga, K.T. 1979. *The Question of Sex Differences.* Boston: Little, Brown.

Hyde, J.S.; Geiringer, E.P.; and Yen, W.M. 1975. On the empirical relation between spatial ability and sex differences in other aspects of cognitive performance. *Multivariate Behavioral Research* 10:289–309.

Kimura, D. 1967. Functional asymmetry of the brain in dichotic listening. *Cortex* 3(2):163–178.

Kinsbourne, M. 1980. If sex differences in brain lateralization exist, they have yet to be discovered. *Behavioral and Brain Sciences* 3:241–242.

Klaus, M., and Kennell, J. 1976, *Maternal-infant bonding.* St. Louis: Mosby.

Knox, C., and Kimura, D. 1970. Cerebral processing of nonverbal sounds in boys and girls. *Neuropsychologia* 8:227–237.

Kreuz, L.E., and Rose, R.M. 1972. Assessment of aggressive behavior and plasma testosterone in a young criminal population. *Psychosomatic Medicine* 34:321–332.

Lamb, M.E. 1975. Physiological mechanisms in the control of maternal behavior in rats: A review. *Psychological Bulletin* 82:104–119.

Lamb, M.E., and Goldberg, W.A. 1980. The father-child relationship: A synthesis of biological, evolutionary, and social perspectives. In R. Gandelman and L. Hoffman (eds.), *Parental Behavior: Its Causes and Consequences*. Hillsdale, N.J.: Lawrence Erlbaum Associates.

Levy-Agresti, J., and Sperry, R.W. 1968. Differential perceptual capacities in major and minor hemispheres. *Proceedings of the National Academy of Sciences of the United States of America* 61:1151.

Maccoby, E.E., and Jacklin, C.N. 1974. *The Psychology of Sex Differences*. Stanford, Stanford University Press.

Macrides, R.; Bartke, A.; and Dalterio, S. 1975. Strange females increase plasma testosterone levels in male mice. *Science* 189:1104–1105.

McEwen, B.S. 1981. Neural gonadal steroid actions. *Science* 24:1303–1311.

McGlone, J. 1980. Sex differences in human brain asymmetry: A cortical survey. *Behavioral and Brain Sciences* 3:215–227.

Mead, M. 1935. *Sex and Temperament in Three Primitive Societies*. New York: Morrow.

Money, J., and Ehrhardt, A.A. 1972. *Man and Woman, Boy and Girl: The Differentiation and Dimorphism of Gender Identity from Conception to Maturity*. Baltimore: Johns Hopkins University Press.

Moss, H.A. 1967. Sex, age, and state as determinants of mother-infant interaction. *Merrill-Palmer Quarterly* 13:19–36.

Nash, S.C. 1979. Sex role as a mediator of intellectual functioning. In M.A. Wittig and A.C. Petersen (eds.), *Sex-Related Differences in Cognitive Functioning*. New York: Academic Press.

Newsweek. 1975. The women's touch. January 6, p. 35.

Parke, R.D., and Sawin, D. 1977. The family in early infancy: Social interactional and attitudinal analyses. Paper presented at Society for Research in Child Development, New Orleans.

Parmalee, A.H., Jr., and Stern, E. 1972. Development of states in infants. In C.D. Clemente, D.P. Purpura, and F.E. Mayer (eds.), *Sleep and the Maturing Nervous System*. New York: Academic Press.

Parsons, J.; Adler, T.F.; Futterman, R.; Goff, S.B.; Kaczala, C.M.; Meece, J.; and Midgley, C. 1981. Expectancies, values and academic behaviors. In J.T. Spence, (ed.), *Assessing Achievement*. San Francisco: Freeman.

Persky, H.; Smith, K.D.; and Basu, G.K. 1971. Relation of psychologic measures of aggression and hostility to testosterone production in man. *Psychosomatic Medicine* 33(3):265–277.

Persky, H.; O'Brien, C.P.; Fine, E.; Howard, W.J.; Khan, M.A.; and Beck, R.W.; 1977. The effect of alcohol and smoking on testosterone function and aggression in chronic alcoholics. *American Journal of Psychiatry* 134:621–625.

Petersen, A.C. 1979. Hormones and cognitive functioning in normal development. In M.A. Wittig and A.C. Petersen (eds.), *Sex-Related Differences in Cognitive Functioning.* New York: Academic Press.

————. 1980. Biopsychosocial processes in the development of sex-related differences. In J.E. Parsons (ed.), *The Psychobiology of Sex Differences and Sex Roles.* Washington: Hemisphere.

Redican, W.K. 1976. Adult male-infant interactions in non-human primates. In M.E. Lamb (ed.), *The Role of the Father in Child Development.* New York: Wiley.

Reinisch, J.M. 1976. Prenatal exposure to synthetic progestins increases potential for aggression in humans. *Science* 211:1171–1173.

Reinisch, J.M.; Gandelman, R.; and Spiegel, F.S. 1979. Prenatal influences in cognitive abilities. In M.A. Wittig and A.C. Petersen (eds.), *Sex-Related Differences in Cognitive Functioning.* New York: Academic Press.

Reiter, R.R. (ed.). 1975. *Toward an Anthropology of Women.* New York: Monthly Review Press.

————. 1976. Unraveling the problem of origins: An anthropological search for feminist theory. In *The Scholar and the Feminist III: Proceedings.* New York: Barnard College.

Riegel, K.F. 1976. The dialectics of human development. *American Psychologist* 31:689–700.

Rosaldo, M.Z., and Lamphere, L. (eds.), *Women, Culture and Society.* Stanford, Stanford University Press.

Rose, R.M.; Holaday, J.W.; and Bernstein, I.S. 1976. Plasma testosterone, dominance rank and aggressive behavior in male rhesus monkeys. *Nature* 231:366–368.

Rosenblatt, P., and Cunningham, M.R. 1976. Sex differences in cross-cultural perspective. In B.B. Lloyd and J. Archer (eds.), *Exploring Sex Differences.* London: Academic Press.

Rossi, A.S. 1977. A biosocial perspective on parenting. *Daedalus* 106(2):1–32.

Rubin, R.T.; Reinisch, J.M.; and Hasbett, R.F. 1981. Postnatal gonadal steroid effects on human behavior. *Science* 211:1318–1324.

Rudel, R.G.; Denckla, M.B.; and Spaltar, E. 1974. The functional asymmetry of Braille letter learning in normal sighted children. *Neurology* 24:733–738.

Sameroff, A.J. 1977. Early influences on development: Fact or fancy. In S. Chess and A. Thomas (eds.), *Annual Progress in Child Psychiatry and Child Development.* New York: Brunner/Mazel, pp. 3–33.

Scheinfeld, A. 1958. The mortality of men and women. *Scientific American* 198:22–27.

Schratz, M. 1978. A developmental investigation of sex differences in spatial (visual-analytic) and mathematical skills in three ethnic groups. *Developmental Psychology* 14:263–267.

Sherman, J.A. 1967. Problems of sex differences in space perception and aspects of intellectual functioning. *Psychological Review.* 74:290–299.

Sherman, J. 1971. *On the Psychology of Women: A Survey of Empirical Studies.* Springfield, Ill.: Thomas.

———. 1980a. Mathematics, spatial visualization, and related factors: Changes in girls and boys, grades 8–11. *Journal of Educational Psychology* 72:476–482.

———. 1980b. Sex-related differences in functional human brain asymmetry: Verbal function—no; spatial function—maybe. *Behavioral and Brain Sciences* 3:248–249.

———. 1980c. Women and mathematics: Summary of research from 1977–1979 NIE grant. Final report to National Institute of Education, Washington, D.C.

———. In press. Girls' and boys' enrollment in theoretical math courses: A longitudinal study. *Psychology of Women Quarterly.*

Sperry, P.W., and Levy, J. 1970. Mental capacities of the disconnected minor hemisphere following commissurotomy. Paper presented at American Psychological Association, Miami.

Stafford, R.E. 1961. Sex differences in spatial visualization as evidence of sex-linked inheritance. *Perceptual and Motor Skills* 13:428.

Starr, B.S. 1979. Sex differences among personality correlates of mathematical ability in high school seniors. *Psychology of Women Quarterly* 4:212–220.

Steel, L., and Wise, L. 1979. Origins of sex differences in high school math achievement and participation. Paper presented at American Educational Research Association, San Francisco.

Waber, D.P. 1979. Cognitive abilities and sex-related variations in the maturation of cerebral cortical functions. In M.A. Wittig and A.C. Petersen (eds.), *Sex-Related Differences in Cognitive Functioning.* New York: Academic Press.

West, M.M., and Konner, M.L. 1976. The role of the father: An anthropological perspective. In M.E. Lamb (ed.), *The Role of the Father in Child Development.* New York: Wiley.

Whiting, B., and Edwards, C.P. 1973. A cross-cultural analysis of sex differences in the behavior of children aged three through 11. *Journal of Social Psychology* 91:171–188.

Wilson, J.D.; George, F.W.; and Griffin, J.E. 1981. The hormonal control of sexual development. *Science* 211:1278-1284.

Williams, T. 1975. Family resemblance in abilities: The Wechster scales. *Behavior Genetics* 5:405-409.

Wise, L.; Steel, L.; and MacDonald, C. 1979. Origins and carrier consequences of sex differences in high school mathematics achievement. Final report to National Institute of Education, Washington, D.C.

Wittig, M.A., and Petersen, A.C. (eds.). 1979. *Sex-Related Differences in Cognitive Functioning.* New York: Academic Press.

Young, W.C.; Goy, R.W.; and Phoenix, C.H. 1964. Hormones and sexual behavior. *Science* 143(3603):212-218.

Zussman, J.U.; Zussman, P.D.; and Dalton, K. 1977. Effects of prenatal progesterone on adolescent cognitive and social development. Paper presented at International Academy of Sex Research, Bloomington, Ind.

Measuring the Impact of Environmental Policies on the Level and Distribution of Earnings

Paul Taubman

With the exception of research on recombinate DNA and perhaps test-tube babies, policy research in this country is concerned with proposing and evaluating various changes in the environment. Such policies try to improve an individual's performance, eliminate harmful behavior, and, in general, overcome poor genetic endowments and family background. These policies operate either by providing services directly or by lowering the price of services. Economists have studied environmental policies that relate to many different subjects. This paper will focus on earnings and its relationship with schooling and with inequality of opportunity. These two subjects will be examined separately.

While most policies studied are environmental in nature, one's knowledge of their impacts may be sorely limited if one ignores or does not control for a person's genetic endowments. Perhaps the simplest way of illustrating this point is in terms of the impact of schooling on earnings. It is often argued that the reason the more educated have higher earnings is that the more educated are more able, irrespective of education, and that ability is rewarded in the marketplace. Thus, not controlling for this ability, which is partly attributable to differences in both family environment and genetic endowments, will cause the researcher to obtain a biased estimate of the effect of schooling on earnings.[1]

The argument can be formalized as the bias that arises when a variable is omitted. Let earnings be denoted by Y, years of schooling by S, ability by A, and random events by u. Let the equation to be estimated be:

$$Y = \beta S + \gamma A + u \tag{9.1}$$

If one omits the variable A, under standard assumptions the expected value of $\hat{b}$, the least squares estimate of the coefficient on schooling is given by:

This research has been supported by NSF Grants SOC76–17673. This piece is based on work done jointly with J. Behrman, T. Wales, and Z. Hrubec. Besides them thanks are due to O. Chamberlain, A. Goldberger, Z. Griliches, C. Jencks, and M. Olnech for helpful criticisms.

171

$$E(\hat{b}) = \beta + \gamma d \qquad (9.2)$$

where d is given by the auxiliary equation

$$A = dS \qquad (9.3)$$

Thus the estimate of β will be biased unless ability does not have an independent effect on earnings ($\gamma = 0$) or unless ability is not linearly related to schooling ($d = 0$).

This well-known omitted variable problem permeates most empirical, nonexperimental research in the social sciences. The two standard approaches to overcoming the problem can best be understood from equations 9.4 and 9.5.

$$A = \lambda IQ + v \qquad (9.4)$$

$$A = \delta_1 X + \delta_2 Z \qquad (9.5)$$

In 9.4 and 9.5, A, v, and Z are unobserved while IQ and X are observed. Equation 9.5 states that there is an observed variable such as an IQ test that is a *direct measure* of ability. Equation 9.4 includes a random variable (v) either because IQ has a test-retest type measurement error and/or because IQ does not correspond exactly to A.

Equation 9.5 asserts that there is a relationship between ability and some observed variables (X) plus some unobserved variables (Z). The relationship in 9.5 of the X variables to A can arise *because X causes A* or *because X is correlated with A*. This latter correlation can occur because of a variety of reasons. For example, some identifiable socioeconomic groups may be willing to divert more resources to their children's future than their own present consumption. Then X would be a proxy for the unobserved differential in resources. Alternatively, some variables such as parents' education may be due to their parents' genetic endowments, which will be correlated with a child's genetic endowments, as explained more fully below. The Z variable includes all the causes of A, including genetic endowments, not correlated with X.

Substituting equations 9.4 and/or 9.5 into 9.1, economists and sociologists have included IQ and various Xs as controls in earnings functions. The results have been mixed.[2] In general IQ and a number of other variables, including parental income and education, and the subject's religious upbringing, are statistically significant. However, in a majority of the studies the estimate of β in equation 9.1 declines by 15 percent or less when such controls are introduced. These studies suggest that there is not a large bias from not controlling for ability. However, in a minority of studies, the bias from not using these controls is 25 to 40 percent, which suggests a large bias if ability is not controlled. I am struck by the fact that the minority group is based on samples in which the men had substantial amounts of labor-force experience while the majority group is based on samples in which the men generally had less than seven years' worth of labor-force experience. However, since nearly all samples with measures of intelligence are based on nonrandom, idiosyncratic samples, there may be other reasons for the difference in results.

While these studies just discussed are valuable, they suffer from at least three important defects. The first problem is that A in equation 9.1 is called *ability*. Many different abilities are rewarded in the marketplace. Very few studies have direct measures of more than a few abilities; and the most widely available proxies of family background, such as parental education, would not seem to be good proxies for many abilities. In other words, the variances of v and Z in equations 9.4 and 9.5 may be relatively large. Of course, a bias on the schooling coefficient will occur only if v or Z is correlated with schooling, that is, if the equivalent of d is nonzero. It is not possible, however, to know a priori if all necessary dimensions of A have been controlled.

The second problem is one of possible misinterpretation. Many studies have found that parental education and income and other aspects of family background are statistically significant and have large coefficients in equations for earnings, schooling, etc. There is a tendency to argue that these coefficients indicate the causal effect of increased parental education, etc., on a child's ability and earnings. These results are completely consistent with a model in which better educated or wealthier parents provide a better environment for their progeny. The results are also consistent with a model in which family background is a proxy solely for genetic endowments of the child. If the latter model is the correct explanation, then family background variables do not have a causal interpretation even if they are adequate proxies to control for ability.[3]

The third problem arises if schooling is measured with error. It is well known that random measurement error in an independent variable will bias the coefficient toward zero. Griliches (1978) shows that the magnitude of the measurement error bias per se increases when measures of A are included in the equation because the size of the bias depends on the ratio of the variances or measurement error to the variance in schooling after partialing out the ability variable.

Controlling A through Twins

In this chapter I will use a different approach to control for ability differences. Under certain conditions examined below, it is possible to use data on identical twins to obtain unbiased estimates of β (in equation 9.1). A more complicated technique that uses data on both identical and fraternal twins can also be used to obtain unbiased estimates and to test some important assumptions that underlie the simpler method.

To understand the argument, a short biological detour is necessary. Genes are located on chromosomes. Each gene at a particular location has two members, one of which is contributed by each parent. The two members can be the same or different, for example, AA or AB. It is known that some genes come in a variety of forms. Each egg and each sperm contain one member, randomly determined, of each gene. When the egg is fertilized by the sperm, the two members combine to form the child's gene. Assume that many genes have small influences on a particular skill. While sibs will receive the same member of one gene from each parent one-fourth of the time, the probability that they will receive the same member from each parent on each gene approaches zero rapidly. In other words, since each egg and each sperm is different, children of the same parents are in general not genetically alike.

There are two twin types, identical (MZ) and fraternal (DZ). Fraternal twins occur when two eggs happen to be released in the same month and each is fertilized by a separate sperm. Fraternal twins are just siblings born at the same time and are not genetically alike.

Identical twins occur when an already fertilized egg happens to split. Both halves of the split egg are genetically alike unless a mutation occurs. Indeed, the definitive way to determine that a twin pair is identical is to demonstrate that they are biochemically alike on a large enough set of factors such as blood type that are solely determined by a person's genetic makeup.

Earlier I noted that genes can come in a variety of combinations. Suppose that I am talking about a particular ability, A_1. For each gene combination that affects A_1 there is associated a particular level of ability that is called the *genotype* (G). A_1 will also be affected by a person's environment, N. G and N combine to produce A_1. The form of the production function is unknown. For the moment, assume that the production function is linear. Since G and N are unobserved, I can express each in units such that

$$A_1 = G_1 + N_1 \qquad (9.6)$$

where I have put subscripts on G and N to indicate that they are the genotype and environment for the first ability.

Classical geneticists generally define or measure G at the mean of all possible environments. From the viewpoint of an economist, it is preferable to assume that parents and their offspring in part choose their environment, and that these choices are conditional on each person's genetic endowments. If this viewpoint is accepted, the G term in equation 9.6 will be defined to include this impact on environment while the N term will be net of this effect. This difference in definition and interpretation is important in this chapter at only a few points, appropriately noted.

Equation 9.6 applies to one type of ability. Equivalent equations perhaps with different G and N apply to all abilities. These various abilities can be combined into an overall ability variable that is related to an overall genotype and environment variable as

$$A = G + N \tag{9.6a}$$

Previously I argued that G would be the same for identical twins in a family but different for fraternal twins in a family. Both types of twins, however, share some environment because they are nurtured in the same womb and generally are raised by the same parents in the same neighborhood till they are at least sixteen years old.[4]

Economists generally think of the environment as investments in human capital though some people may think that such a term is not a felicitous relabeling when applied to love, affection, and time spent by parents with their children. A standard investment in a human capital model for the person f in family k would be 9.7[5]

$$I_{jk} = aP_{jk} + cr_{jk} + eY_{jk} + mG_{jk} + nT_{jk} + w_{jk} \tag{9.7}$$

where I is investment, P is the market price of investment, r is the interest rate on borrowed funds, Y is family income, T is family tastes, and w is a random variable.

The same equation could be written for person j's twin. Then I_{1k} and I_{2k} will be correlated because the variables on the right-hand side of 9.7 will also be correlated for twins. Indeed, it seems likely, if the investments that influence ability and whose omission biases the estimate of β are undertaken during childhood, that P, r, and Y will be the same for both twins reared together in family k. Within family k, G and T will also be correlated while w will not be correlated. Again I remind the reader that the unobserved N in equation 9.6 or 9.6a equals $I_{jk} - mG_{jk}$.

I can, of course, easily measure the one dimension of environment called *years of schooling*. Once I separate out schooling from other unobserved environments, formally I should relabel the G and N in equation 9.6; however, to simplify notation I will not do so. My model can be stated as equations 9.1, 9.6a, and 9.7.

Randomly order pairs within a family, and denote within pair differences by a Δ. ΔY_k, for example, is $Y_{1k} - Y_{2k}$. Then after substituting 9.6a into 9.1, I can write

$$\Delta Y_k = \beta \Delta S_k + \gamma (\Delta G_k + \Delta N_k) + \Delta u \tag{9.8}$$

For identical twins, ΔG_k is always zero, and from 9.7 N_k reduces to $(n\Delta T_k + \Delta w_k)$. Ordinary least squares applied to 9.8 will yield an unbiased estimate of β, provided that (1) ΔS_k is not correlated with $(n\Delta T_k + \Delta w_k)$; (2) S is measured without error; and (3) ΔS_k is not zero for some k.

I will consider the importance of these three provisos in a moment. First, however, I wish to note that if I apply equation 9.8 to fraternal twins, the ΔG_k terms are not in general equal to zero. Thus ΔG_k is an omitted variable that can cause a bias. Griliches (1978) elegantly demonstrates that under certain conditions the within-pair regressions for fraternal twins may yield a more biased estimate than that obtained from equation 9.1.

Now let us return to the three provisos. In my sample about one-half of the identical twin pairs do not report the same years of schooling, though most of the discordant pairs differ only by a year or two. As will become evident, this is a large enough sample to estimate the equation.

Measurement error in schooling is a more important problem. Measurement error causes a much greater bias in within-pair equations than in corresponding equations using data on individuals.[6] Thus far I have not been able to determine exactly the measurement error in schooling in my sample. A variety of sources allows me to guess at the magnitude. For example, I have estimates of parents' education from both twins. In any event, I can calculate the extent of the bias for any assumed variance in measurement error.

The third proviso was that ΔS was uncorrelated with ΔN. The major reason for this correlation to be nonzero is that parents consciously allocate more of many types of resources to one or the other of the siblings. That is, parents may like one of the twins more than the other and invest more in him than in his twin. Alternatively, one twin may be thought to be able to make more out of the investments because he is more industrious or has not suffered a debilitating accident.

It is possible to test the proposition that for identical twins ΔS is not correlated with ΔA.[7] Suppose that parents wish to provide more investments of all types of human capital to one sib in an MZ twin pair. Then the value of the unobserved ability variable in equation 9.1 will be higher for this child. If I order the children within a pair by schooling level, the genetically identical child with more schooling would also have more unobserved investments and ability. Thus the within-pair equations ordered by schooling should have a positive constant term.

The second test is much more complicated. It requires setting up a recursive model whose dependent variables are schooling, initial and mature occupational status, and mature earnings. By imposing a number of restrictions on this model, it is possible to test if differences in the noncommon environment that affect differences in earnings directly are correlated with differences in schooling. It is in this model that we make use of the alternative definition of genotype referred to earlier.

There is an additional reason why MZ within-pair equations may not yield unbiased estimates of β. Equation 9.8 is a combination of equations 9.1 and 9.6 or 9.6a. Equation 9.6 is a linear production function. It is possible that a nonlinear specification such as

$$Y = G + N + (NG)^{\beta} \tag{9.6b}$$

is more appropriate. If 9.6b is correct, MZ within-pair equations would contain $(G\Delta N)^{\beta}$ terms that are not zero. Jinks and Fulker (1970) developed a test of the null hypothesis that 9.6a is valid. The essence of the test is that if 9.6b is valid then the variance of MZ within-pair differences in earnings will include $(G\Delta N)^{2\theta}$ terms whose magnitude will vary with G. In other words, the error in the earnings equation will not be homoskedastic. The average earnings of a pair is an imperfect measure of the pairs G. A regression of the MZ absolute differences in Y or the square root of the $(\Delta Y_k)^2$ on $\overline{Y}_k$ is a test for heteroskedasticity. For earnings I find substantial heteroskedasticity but for the ln of earnings, I find homoskedasticity. Thus in our empirical work I will use the ln of earnings.

The Sample

The twin sample I use is described in detail in Behrman et al. (1980) and in Taubman (1976). Briefly, in 1955 the National Academy of Science–National Research Council (NAS–NRC) started a project to construct a twin panel as a resource in studying a variety of questions. The NAS–NRC initially collected birth certificates on nearly all white male twins born between 1917–1927. The twin panel consists of those pairs both of whose members are veterans. My sample consists of those pairs in the panel who answered a 1974 survey. Even this brief description should indicate that the sample need not be a random draw from the population of twins. In addition there are a variety of reasons why twins need not be a random draw of the total population. For example, twins are not reared in a single-sib household.

Compared to white males of the same age cohort or to the veteran subsample, the respondents in our sample have above-average earnings. However, my earnings equations yield very similar coefficients to those obtained from random samples of the population drawn in the same time period. I am working on the issue of how generalizable my results are to the population of white males. In the meantime I can only hope that they apply to the larger group.

The sample consists of roughly 1,000 pairs of identical twins and 900 pairs of fraternal twins. For those pairs where both served in the Navy as enlisted men, the person's score on the General Classification Test is also available.

For both the within-pair analysis and the analysis of variance to be described below, it is necessary to distinguish identical and fraternal twin types. In principle this can be accomplished by using a large enough battery of biochemical tests since a pair of identical twins must have the same blood type, RH factor, etc. on all such tests. Such tests have been done for a minority of the pairs. For the most part, however, the twins have been classified on the basis of their answers to the following question: "In childhood were you as alike as peas in a pod or only of ordinary family resemblance?" This may seem like a very delicate item on which to base our analysis, but Jablon et al. (1967) have shown that for this sample, peas in a pod are 95 percent as accurate as biochemical tests. I might add that I was brought up on canned peas, and I do not know how alike peas in a pod are. Indeed I have been told that there is a minor crisis in twin research since many more children and younger adults are as gastronomically deprived as I was and currently respond incorrectly to this question.

The Effect of Schooling on Earnings

In a large number of studies in economics, the semilog specification given in equation 9.1, amended to include years of work experience, has been used. Table 9-1 contains a summary of some sample regression results calculated across individuals and within pairs for 1973 earnings, when the men were about 50 years old. When as in line 1 I treat all the individuals in the sample as unrelated individuals, and when I control for no other variables, the coefficient on years of schooling is a highly significant .080, which is unchanged when age is added. This equation's estimated coefficients on years of schooling are similar to those obtained from nationwide random samples collected in 1960 or later (see for example Lillard and Willis 1978 or Jencks et al. 1972). From line 2 we see that holding constant a whole host of observed family background variables and marital status reduces the coefficient by about 12 percent. For those pairs where both were in the Navy, I have data on the General Classification Test, which is primarily a vocabulary test and a measure of cognitive skills. Controlling for this test and for the variables in the previous line, and estimating the equation across pair averages reduces the education coefficient to .051, a reduction from line 1 of about 35 percent and from line 2 of about 25 percent.[8]

In line 4 the estimated coefficient from the within-DZ-pair equation is a highly significant .059. This is about 25 percent less than the estimate in line 1 and intermediate between line 2, in which background is controlled, and line 3, in which background plus IQ are controlled. Line 5 presents the

Table 9-1

Coefficient on Years of Schooling in Equation for 1n of 1973 Earnings

Equation Based on	Number of Observations	Coefficient on Yrs./Schooling	t Statistic on Coefficient	Other Variables Held Constant
1. Individuals	3870	.080	32.4	none
2. Individuals	3870	.069	25.8	A, B
3. Navy pairs	404	.051	6.0	A, B, C
4. Within DZ pairs	914	.059	8.3	B
5. Within MZ pairs	1022	.026	3.5	B

Other variables held constant, A is: age, number of sibs alive 1940, father's years of schooling, mother's years of schooling, father's occupational status (Duncan score); and the following variables coded as (0,1 dummies) raised in rural area, raised as a Catholic, raised as a Jew, born in the South.

B: married in 1974

C: score on Navy General Classification Test

within-MZ-pairs results. Here the coefficient on schooling plummets to (a still statistically significant) .026, which is one-third of the estimate in line 1 and one-half of the estimate in line 3. Unfortunately this .026 is almost surely biased toward zero by measurement error.

As shown in note 6, the magnitude of the bias arising from measurement error depends on three elements: the ratio of the variance of the measurement error to the variance of true schooling (σ_w^2/σ_s^2); the cross-twin correlation in true schooling (ϱ_s), and the cross-twin correlation in measurement error (ϱ_w). If I assume that ϱ_w is zero, then I can calculate the measurement error bias for any assumed value of σ_w^2/σ_s^2. If this ratio of the variance is about 20 percent then both the MZ within-pair and the individual equations would yield an unbiased estimate of b of about .09. If, however, this ratio is 10 percent, the corrected estimate from the within-pair and individual equations is about 0.044 and .088, respectively.

The one piece of evidence on the size of the measurement error in education in this sample comes from the twins' reports on their parents' education. From these independent reports, I can calculate the average σ_w^2/σ_s^2 of the parents' education to be slightly less than 10 percent. It seems likely that each twin knows his own education more accurately than his parents' education. Thus 10 percent seems a likely upper bound.

As noted above, for this magnitude of measurement error, the .026 figure should be adjusted to about .045. While .045 is nearly 100 percent greater than the estimate in line 5, it is also about 50 percent of .088, the line 1 estimate adjusted for the same degree of measurement error. Thus in studying the effects of schooling on earnings in this sample, it is crucial to control for genetic endowments and family endowment. It also appears that measures of cognitive skill and certain aspects of family background provide fairly adequate controls in that the estimate in line 3 is .05, which is quite close to .044. However, the estimate of .05 in line 3 would also be biased toward zero by measurement error. Thus there must be other abilities omitted from equation 9.3, with the biases from these omitted variables and measurement error approximately offsetting one another.

Based on these results, it appears that identical twins provide the best controls for studying the returns to education. However, a good measure of cognitive skills and a battery of background variables do nearly as well. Since it is probably easier and cheaper to design and collect random samples of individuals than samples of twins with the latter data, this is not a trivial conclusion.

Inequality of Opportunity

While identical twins can be used to control for unmeasured ability, by far the most common use of twin samples has been in nature-nurture studies. Using equation 9.6a, I can express the observed variance in earnings in terms of the unobserved G and N variables as:

$$\sigma_Y^2 = \sigma_G^2 + \sigma_N^2 + 2\sigma_{GN} \tag{9.9}$$

Dividing through by σ_Y^2, I have

$$1 = \frac{\sigma_G^2}{\sigma_Y^2} + \frac{\sigma_N^2}{\sigma_Y^2} + 2\frac{\sigma_{GN}}{\sigma_Y^2} \tag{9.9a}$$

If σ_{GN} is zero, the first term on the right-hand side of 9.9a is the contribution of nature to the variance in Y while the second term is the contribution of nurture. If σ_{GN} is not zero, there is no nonarbitrary definition of the contribution of nature and of nurture.

As the reader may be aware, there has been an extensive and at times bitter debate in the literature on IQ over the nature-nurture question. This debate has focused both on the statistical question of the properties of the estimates of σ_G^2/σ_Y^2 etc. and on the interpretation or implication of the results. It is the implication question that is, I believe, the basis of the bitterness. Rather than trying to summarize the huge literature on the subject, let me give an interpretation.

An analogy may put this question into perspective. It is known that hemophilia is caused by a genetic defect that prevents the blood from clotting. Several years back a drug was developed that will allow clotting to occur. Thus this genetic disease is curable by a change in the environment. But when this drug was introduced it was very expensive—as I recall, something like $20,000 per year per person. Few families now can afford such an expenditure. Society, of course, can choose to pick up the bill. However, the taxes raised to pay for the drug will distort incentives in the economy and may cause individuals to work and save less. The changes will lower the value of goods and services produced now and in the future. This reduction is called *economic inefficiency.*

Of course some genetic defects can be overcome at a much lower cost. For example, some eye diseases may be genetic in origin and the cure may consist of a pair of eyeglasses. In general, in deciding on the undertaking and scope of various policies, society compares the benefits and the costs. Economic inefficiency is one of the relevant costs.

It is, of course, possible to redistribute income or earnings through transfer schemes or wage subsidies. Many people argue, however, that meaningful redistribution involves too great a loss in economic efficiency. This inefficiency occurs for two reasons. The first is that taxes must be raised to pay for this redistribution and, as argued above, such taxes distort labor-leisure and consumption-saving choices. The second reason for inefficiency is that most transfer schemes provide incentives for people to avoid work. If, however, inequality in earnings arises because of inequality of opportunity (defined rigorously below), then it is possible to redistribute earnings while increasing economic efficiency. On precisely these grounds, Okun (1975) voices the hope that inequality of opportunity is a major source of inequality of earnings. Inequality of opportunity, as used by Okun and by most other economists, means that some people are not able to invest in skill acquisition as much as others because of differences in prices and income or perhaps because of discrimination.

The N in equation 9.6a can be split into common and specific environments. Twins share some common environment, which differs across families. The difference across families is, I believe, most due to differences in family income and preferences. The contribution of common environment can be used to estimate the importance of inequality of opportunity.

The model that I use to obtain my estimate is given in table 9-2. Technically I am using a latent-variable, variance-components model .[9] The key assumption that I make to partition the variance is that the expected value of the cross-twin correlation in environment is the same for identical and fraternal twins. If this assumption is not made, it is possible to estimate the contribution of G to be zero.[10] However, the major reason advanced for this correlation to be greater for the identical twins is that parents and the

Table 9–2
Basic Model with $\varrho^* = \varrho' = 1.0$, $\lambda \neq 1/2$ (24 parameters)

	G_1	N_1	G_2	G_3	G_4	u_1	u_2	u_3	u_4	S	OC_i	OC_{67}
Reduced form equations												
S	1.85 (15.9)	1.98 (17.5)				1						
OC_i	.68 (5.5)	1.16 (11.7)	1.16 (18.5)			.21 (6.0)	1					
OC_{67}	.69 (6.3)	.78 (8.4)	.37 (5.1)	.82 (13.8)		.29 (8.9)	.14 (5.4)	1				
$\ln Y_{73}$	.17 (5.9)	.19 (7.7)	.098 (6.0)	.019 (.8)	.31 (26.2)	.026 (3.4)	.0044 (3.5)	.031 (4.7)	1			
Structural equations												
S	1.85 (15.9)	1.98 (17.5)				1						
OC_i	.30 (1.9)	.75 (5.7)	1.16 (18.5)				1			.21 (6.0)		
OC_{67}	.113 (.1)	.090 (1.0)	.20 (2.3)	.82 (13.8)				1		.26 (7.9)	.14 (5.4)	
$\ln Y_{73}$	.12 (3.3)	.13 (5.2)	.087 (5.2)	−.0068 (.3)	.31 (26.2)				1	.016 (2.3)		.031 (4.7)

Other estimates

λ = .34 (6.1)

$\sigma^2_{u_1}$ = 2.17 (22.6)

$\sigma^2_{u_2}$ = 2.75 (24.4)

$\sigma_{u_3}^2 = 2.45$
(25.1)

$\sigma_{u_4}^2 = .127$
(23.1)

Normalizations and restrictions

A,B,C,D,E,F

Functional value: $+13431.87$

Restrictions and Normalizations

A is $\sigma_{G_1}^2 = \sigma_{G_2}^2 = \sigma_{G_3}^2 = \sigma_{G_4}^2 = \sigma_N^2 = 1$

B is $\sigma_{N_2}^2 = \sigma_{N_3}^2 = \sigma_{N_4}^2 = 0$

C is $\sigma_{N_1 G_1} = 0$

D is $\sigma_{N_1 G_1} = 0, i = 2,4$

E is $\sigma_{N_i N_j} = \sigma_{G_i G_j} = 0, i = 1 \ldots 4, j = 1 \ldots 4$

F is $\varrho' = \varrho^* = 1$

G is $\lambda_1 = 1/2, i = 1 \ldots 4$

The figures in parentheses underneath the point estimates are absolute values of ratios of parameter estimates to estimated asymptotic standard errors.

S is years of schooling.

OC_i is initial full time civilian occupational status, Duncan scale.

OC_{67} is occupational status in 1967, Duncan scale.

lnY_{73} is the logarithm of 1973 earnings.

The actual likelihood function values are given by a functional value noted in the table, aside from a constant.

Table 9-3
Sources of Variances of Schooling, Initial and Later Occupational Status, and Earnings, Basic Model

Percent of Total Arising from	S	OC_i	OC_{67}	$\ln Y_{73}$
$\sigma^2_{G_1}$	36%	08%	11%	10%
$\sigma^2_{G_2}$		23	03	03
$\sigma^2_{G_3}$			15	a
$\sigma^2_{G_4}$				32
$\sum \sigma^2_{G_i}$	36	31	29	45
$\sigma^2_{N_1}$	41	22	13	12
$\sigma^2_{u_1}$	23	02	04	01
$\sigma^2_{u_2}$		46	01	a
$\sigma^2_{u_3}$			53	01
$\sigma^2_{u_4}$				42

Source: Table 9-1.
Totals may not add to 100 percent because of rounding.
aImplies less than 0.5 percent.

twins base choices on each child's genetic endowments. Since I define this to be an effect of genetic endowments, this is not a compelling criticism when I estimate the importance of inequality of opportunity.

Table 9-3 presents the total (the sum of direct and indirect) contributions of the various genetic and environmental variables to each of the four dependent variables. As shown in the table, common environment accounts for 12 percent and 41 percent of the variance of the $\ln$ of 1973 earnings and of schooling, respectively. Presumably, parents have much greater impact over schooling decisions of their offspring than over many postschooling decisions. However, since schooling has only a small contribution to the variance of earnings, the relative size of these two findings is not surprising.

The contribution of across-family inequality of opportunity to the variance in earnings of any other variable is given by the common environment term. The results in table 9-3 suggest that inequality of opportunity has a substantial impact—44 percent—on the variance of schooling, one type of investment in human capital, but only a very modest impact—12 percent—on the variance in earnings. The results taken at face value would indicate that even if it had been possible to eliminate all the variances in family environment when this cohort was being reared, the inequality in earnings would not have been reduced greatly. There are, however, a variety of qualifications and modifications that must be considered.

These estimates come from a model in which it is assumed that genotype and environment are uncorrelated. In my model common environment or investment in human capital depends on family income. Thus it may appear that this assumption is not tenable in a model that examines earnings. Of course family income need not be the same as the parents' earnings genotype, because in this sample many mothers would not have worked after the twins were born, because of inherited wealth, and because parents' earnings would depend on their environments. Still I would expect the child's genotype and environment to be correlated. In my model this correlation is identified in the statistical sense, but I have never been able to get estimates that converge. The way the model is structured, any nonzero covariance will cause a change only in the estimate of common environment. The greater is the genotype, environment covariance, the smaller would be my estimate of common environment. Thus by assuming this correlation is zero, I am attributing to environment all the effect of the genotype-common environment correlation. It could be argued that the impact of the correlation should be shared between genetic endowments and common environment, and that on these grounds I am overstating the contribution of common environment.

Measurement error usually refers to wrong numbers being written down. Transitory income refers to correct numbers for annual earnings being recorded but annual earnings deviating from the proper concept of normal, permanent, or lifetime earnings. Thus transitory income can be thought of as another type of measurement error.

Using the Michigan Panel of Income Dynamics, Lillard and Willis (1978) estimate that for white males, permanent income constitutes about 80 percent of the variance in annual earnings. Their remainder would include both transitory effects and measurement error. Several considerations suggest that the permanent component of the variation in income may be about 90 percent of the total.[11] Adjusting for measurement error and transitory income, common environment would account for about 15 percent instead of 12 percent of the variance in earnings.[12]

A different sort of concern involves discrimination. Because of discrimination, some people would be denied access to certain positions or training programs and/or receive a wage less than their marginal product. Thus discrimination would appear to meet my definition of inequality of opportunity, though discrimination may be directed at groups with different genotypes. My sample contains only white males. Even within this group, there may have been discrimination against, say, Catholics and Jews. If such discrimination does not change the cross-twin correlation coefficients because, for example, earnings of all affected pairs are changed equally, my estimate of common environment would include some of the effects of discrimination. Still it may be true that for this reason my estimate is too small for the whole labor force.

In this section I provide an estimate of the contribution of inequality of opportunity, defined as the variation in common environment, to the variance in earnings. I estimate inequality of opportunity to account for less than 20 percent of the variance in outcomes, but that who one's parents are accounts for about 60 percent of the variance in outcomes. These estimates are based on one sample that is not a random drawing of the population and that covers only one part of the life cycle. Still, if these findings are confirmed in other studies, the implication is clear. Those who wish to reduce greatly the degree of inequality will have to use compensatory training and transfer programs. With these policies the trade-off between efficiency and equality will have to be faced.

Conclusions

For the past several decades much of the research in the social sciences on the sources of individual differences in schooling, earnings, and many other variables has been based on models in which differences in the environment play the major role. Many people have estimated models in which observed portions of the environment such as schooling are entered as explanatory variables. Often the presumption is that the estimated coefficients on these environmental variables indicate the source or cause of the difference in earnings or IQ. The results obtained from the within-pair equations for identical twins suggest that this environmental interpretation is overstated. That is, for men aged about 50, I find that 50 percent or more of the coefficient on schooling obtained in an equation estimated for individuals is actually attributable to statistically uncontrolled differences in ability. Since the bias is much smaller when I use within-pair equations for fraternal twins, it is genetic endowments that underlie the differences in this ability. The results from the Navy subsample indicate that the ability that should be controlled in earnings equations includes cognitive skills but there are other dimensions of ability that are correlated with religion and other aspects of family background. (See notes to table 9–1).

Many studies have found that measures of family background such as parental education are significant in equations for schooling, earnings, etc. Some people interpret these results as indicating that changes in education will have important intergenerational consequences. My results on the partitioning of the variance of earnings into genetic, common, and noncommon environment components suggest that much more of the observed differences in outcome attributed to parents occur because of genetic inheritance than because of differences in family income or environment. While this statement is consistent with the proposition that a change in average family income of $1 will have a large or small effect on average children's earnings, it suggests that the above intergenerational conclusion may be incorrect.

The conclusion that diffences in family income (or common environment) have only a modest impact on the inequality of earnings has another important implication. Suppose that individual's access to capital markets and investments in human capital are limited because of family wealth. Society could eliminate or reduce the importance of family wealth by following a policy of guaranteeing loans or by giving subsidies. The return to society of the elimination of market imperfections can be greater than the social cost of making these loans. Thus elimination of inequality of opportunity can lead to increases in equity and economic efficiency. All other income redistribution mechanisms gain increased equity by sacrificing efficiency. The results from the NAS-NRC twin sample suggest that inequality of opportunity is not a major explanation of inequality of earnings. Society, of course, can still redistribute income to overcome differences in innate ability. The costs of this redistribution will be higher than if inequality of opportunity were important.

Notes

1. A biased estimate is one whose expected value does not equal its true value.

2. For a recent survey, see chapter 1 of Behrman et al. (1980).

3. For a very interesting study on this issue, see Scarr and Weinberg (1977).

4. Some twins are separated. For an excellent example of research with this type of sample, see Shields (1962), who also summarizes other studies.

5. The model can be modified to include parental love, affection, and time. In the following discussion they can be considered as part of the T variable.

6. Let true schooling be s and measured schooling be S. The measurement error is v. That is, $S = s + v$. The true equation is $Y = \beta s + u_2$ but using ordinary least squares our estimate of β will be $\Sigma SY/\Sigma S^2$, whose expected value is $\beta/(1 + \sigma_v^2/\sigma_s^2)$. The corresponding estimate from within-pair equations reduces to $\sigma/(1 + \sigma_w^2(1 - \varrho_w)/\sigma_w^2(1 - \varrho_s))$ where ϱ is the cross-twin correlation coefficient. For example, $\varrho_s = \sigma s_1 s_2/\sigma s_1 \sigma s_2$. We

would expect ϱ_w to be zero and ϱ_s to be positive and large. As long as $\varrho_w <$ ϱ_s, the measurement error bias from the within-pair estimate will be greater than that obtained from individuals.

7. The following tests are taken from Behrman, Pollak, and Taubman (forthcoming).

8. The Navy only has test scores for the enlisted men. To correct for selectivity bias a dummy variable with a one for officer was included.

9. For a complete description, see Behrman et al. (1980), chapter 5.

10. See Behrman et al. (1980).

11. Their 80 percent figure should be a lower bound for my sample for two reasons. First, their sample includes young people, whose earnings grow most rapidly and irregularly, while my sample includes only men about age 50. The rapid growth of the young counts as transitory income in their methodology. Second, I would expect that transitory income of brothers would be correlated because of choices of similar occupations, on-the-job training, etc.

12. If the permanent component is 20 percent, the estimate for common environment would rise to 17 percent.

References

Behrman, J., P. Taubman, T. Wales, and Z. Hrubec. 1980. *Socioeconomic Success: A Study of the Effects of Genetic Endowments, Family Environment and Schooling.* Amsterdam: North-Holland.

Griliches, Z. 1978. A partial survey of sibling models. Cambridge, Mass.: Harvard University, mimeo.

Jablon, S., et al. 1967. The NAS–NRC Twin Panel: Methods of construction of the panel, zygosity diagnosis and the proposed use. *American Journal of Human Genetics* 19:133–161.

Jencks, C., et al. 1972. *Inequality: A Reassessment of the Effects of the Family and Schooling in America.* New York: Basic Books.

Jinks, J.L., and D.W. Fulker. 1970. Comparison of the biometrical, genetical, MAVA, and classical approaches to the analysis of human behavior. *Psychological Bulletin,* 73(May):311–349.

Lillard, L., and R. Willis. 1978. Dynamic aspects of earnings mobility. *Econometrica* (September).

Okun, A. 1975. *Equality and Efficiency: The Big Trade Off.* Washington, D.C.: The Brookings Institution.

Scarr, S., and R. Weinberg. 1977. Intellectual similarities within families of both adopted and biological children *Intelligence* 1:170–191.

Shields, J. 1962. *Monozygotic Twins Brought Up Together and Apart.* Oxford: Oxford University Press.

Taubman, P. 1976. The determinants of earnings, genetics, family and other environments: A study of white male twins. *American Economic Review* (December).

10 Delinquency and Crime: A Biopsychosocial Theory

Juan B. Cortés

A comprehensive theoretical formulation of criminal behavior should (1) organize all the heterogeneous lines of empirical knowledge accumulated by research and investigations; (2) furnish the ultimate reason for why a particular individual adopts criminal patterns and behavior while another one, under similar environmental variables, does not; and (3) offer new hypotheses that in turn may lead to relevant investigations in further fulfillment or modification of the theory. These three objectives will be developed in this chapter.

Factors Related to Criminal Behavior

Many factors are involved in the complex problem of delinquency and crime. While still others could and probably should be included, the clusters of variables listed in figure 10-1 seem among the most important and relevant. It is clear that both biopsychological and sociocultural factors play a critical role. Factors 1-3 of the figure are mostly biological and psychological, whereas Factors 4-12 are mostly cultural and sociological. However, the important aspect that needs emphasis is the mutual interaction among all those factors. The answer to the problem of crime and delinquency must lie in the interaction between the particular personality and the particular environment, or in the *particular reaction of some persons to the elements in their environments.* Most of the variables or factors that affect either the personality or the environment, or both, are presented in figure 10-1 and will be developed more fully in the following subsections.

1. *Mesomorphic physique.* Sheldon (1949) found 64.5 percent mesomorphs among his delinquents versus only 40 percent among his nondelinquents; Glueck and Glueck (1950, 1956) 60.1 percent among their delinquents and 30.7 percent among their nondelinquents; Cortés and Gatti (1972), 57 percent mesomorphs among their delinquents versus 19 percent among the nondelinquents ($X^2 = 19.29$, $p < .0001$), and 80 percent mesomorphs among their sample of criminals. In his excellent study dealing with over 4,000 high-school students, Hirschi (1969) reported: "both white and negro boys who see themselves as 'well-built' (as opposed to fat, skinny or just average), are most likely to have committed delinquent acts" (p.

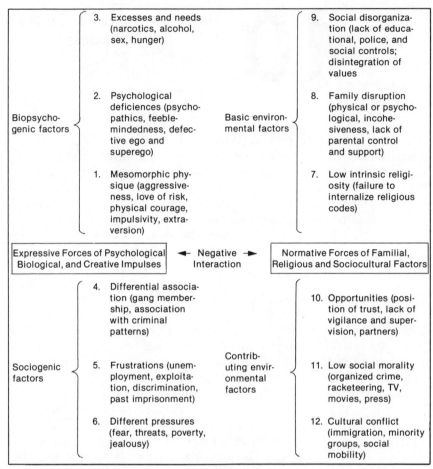

Figure 10-1. Factors Associated with Delinquent and Criminal Behavior

196); see also Gibbens (1963), Lefkowitz and Cannon (1966), West (1969), and Cartwright (1974).

This factor is relevant for several reasons. Mesomorphy seems to be associated with potential aggressivity (Sheldon 1949; Cortés and Gatti 1972) and high need for achievement and power; mesomorphs describe themselves as energetic, assertive, reckless, extroverted of action, impulsive, all of which are temperamental traits found more often among delinquents than nondelinquents (Glueck and Glueck 1950; Cole and Hall 1966; Horrocks and Gottfried 1966; Conger, Miller, and Walsmith 1970).

2. *Psychological deficiencies.* Included under this heading are various psychological and neurological abnormalities and subnormalities such as defective ego and superego (Reiss 1952; Grossbard 1962), feeblemindedness

(Messinger and Apfelberg 1960; Ferracuti 1966), numerous emotional disturbances (Redl and Wineman 1952; Lewis and Yarnell 1951), psychosomatic disorders (Glueck and Glueck 1950; Abrahamsen 1960; Michaels 1961), abnormal electroencephalogram patterns and signs of cortical immaturity (Craft 1966; Lykken 1968; Hare 1968, 1970; Hare and Cox 1978), psychoses (Radzinowicz and Turner 1944; Messinger and Apfelberg 1960). All these factors have been found to be associated in varying degrees with criminality. It is true that in recent years modern thinking has changed and the evidence suggests that the great majority of delinquents and criminals are neither neurotic nor psychotic (see, among others, Yochelson and Samenow vol. I, 1976). However, that delinquents and criminals do have very deficient superegos, as Freud understood it, is definitely a well-established fact. This is even more true concerning psychopaths, the worst category of criminals (Cleckley 1976; Reid 1978).

3. *Excesses and needs.* Some reviewers of Cortés and Gatti (1972) (see Walters 1973) have indicated that clusters of factors such as this one are vague and impossible to research. This, however, runs contrary to the evidence. These excesses and needs include: addiction to narcotics (Maurer and Vogel 1967; Lindesmith 1978; Blum 1969), addiction to alcohol (Shupe 1954; Pittman 1967; Reid 1978), and other physiological needs such as hunger (Tappan 1960) and sex (Radzinowicz 1957; Karpman 1956; Gebhard et al. 1965; Haskell and Yablonsky 1978). Furthermore, it seems obvious that these needs and excesses have been abundantly researched. They tend to be deviant or delinquent in themselves and are likely to lead to other forms of antisocial behavior.

All the preceding factors are referred to as *biopsychogenic* because, particularly compared to the other factors, they appear to be in the particular makeup of the personality. That is, they seem to have a predominantly biological or psychological basis. Whatever their environment, individuals do vary in their mesomorphy, need for sex, amount of extroversion, or proclivity to narcotics and other drugs. All these factors may weaken self-control and may thus facilitate the manifestations of instinctual or expressive impulses. The reader should keep in mind, however, that environmental elements also have some bearing upon all these clusters of factors.

4. *Differential association.* I am very much opposed to this factor as a *general theory* of delinquency and crime (Cortés and Gatti 1972, pp. 164–178). Even Sutherland rejected his own theory and declared it invalid (Sutherland in Cohen et al. 1956). However, association with others manifesting criminal patterns of behavior does certainly play an important role in many instances of criminal behavior. Gangs (Yablonsky 1970; Robin 1964; Thrasher 1963; Cohen 1966; Spergel 1967; Short 1968; Miller 1970), interaction with criminals (Glueck and Glueck 1943), delinquent friends (Hirschi 1969), unwholesome companionships (Cloward and Ohlin 1960; Empey 1978); subcultures of violence (Wolfgang and Ferracuti 1967), and delin-

quency areas (Shaw and McKay 1969) are undoubtedly factors relevant to criminal behavior.

5. *Frustrations.* Many external or environmental elements can lead to frustrations of the type included here, which, in turn, may lead to criminal behavior (Henry and Short 1954). Among them: exploitation (Sutherland 1961), unemployment (Winslow 1939; Clark 1970), underemployment (Kerner Report 1968), past imprisonment (Glueck and Glueck 1943), racial conflicts (Report on Violence 1970), an assassination of such a leader as Martin Luther King, discrimination, inadequacy of federal programs, police practices (Kerner Report 1968), and many others.

6. *Various pressures.* Not a small number of crimes are committed out of fear or threats, because of jealousy, blackmail, as a last resort (Sutherland in Cohen et al. 1956; Tappan 1960), or due to all kinds of economic pressures (Berg 1967), poverty (Clark 1970; Report on Violence 1970), or, perhaps more accurately, to poverty in the midst of affluence (Toby 1967).

The factors included in clusters 4–6 are labeled *sociogenic* because they seem to proceed mainly from particular circumstances found in society or in the subcultures and individuals or elements immediately surrounding the potential offender. If we consider criminal behavior in terms of an excitation-inhibition imbalance (Eysenck 1957, 1970), all the clusters of factors enumerated so far seem to have more effect on the excitation aspect of such imbalance. They operate on the expressive forces of the individual personality by provoking, arousing, and exciting them. However, criminality may also be produced by weakening the inhibitions, loosening social controls, diminishing internal resistance and the self-concept, thereby facilitating nonconforming reactions to elements, opportunities, and situations of the environment. In the following subsections, clusters of factors will be outlined that in a more direct sense or to a greater degree, influence the inhibition aspect of the excitation-inhibition imbalance by lowering or failing to establish it.

7. *Low intrinsic religiousness.* I have studied this variable and found a consistent and extremely significant difference in religiousness between nondelinquent and delinquent boys and between the parents of nondelinquent and delinquent boys. Of all the 55 factors we studied (Cortés and Gatti 1972), this was the most discriminating one. There seems to be little doubt that this variable is relevant to the problem of delinquency and criminality (see Argyle 1959; Fitzpatrick 1967; Rosenquist and Megargee 1969). However, it has not been determined whether low religiousness is a result of, or a predisposition to, delinquent behavior. Very likely it is both and affords some circularity to the model: low religiousness would lead to delinquency, and delinquent behavior would lower religiousness.

8. *Family disruption.* The evidence for this cluster of variables being a relevant factor in antisocial behavior is very conclusive. Family disruption is

understood here both as physical disruption (broken homes due to divorce, separation, or desertion) and psychological disruption (tensions, conflicts, even hate, between parents and children or between husbands and wives). Delinquents come from broken homes more frequently than do nondelinquents, as shown in numerous studies (Healy and Bronner 1936; Monahan 1957, 1960; Pati 1961; Rodman and Grams 1967), and they receive less parental control and support (Glueck and Glueck 1950, 1962; Cortés and Gatti 1972). Abrahamsen (1960, p. 311) has written, "a common factor in all delinquency studies is dissatisfaction with their parents." Wilkins (1960) and Biller (1970) stress the negative impact of father absence; Hirschi (1969), low attachment to parents; and Toby (1957) and Nye (1958), family disorganization. Chateau (1961) concluded that lack of paternal authority and control is a major factor in juvenile misconduct, and Zamorano (1961) stated that well-organized family life is of very great importance for the prevention of delinquency.

9. *Social disorganization.* One phase of the sociocultural context that is especially relevant to the incidence of crime is the disorganization of the normative system by which behavior is regulated. The Reiss (1951) study on delinquency as a failure of social controls is pertinent here. Lack of social organization, disintegration of values, riots, police strikes, deficiencies in police control, school problems and inadequacies, and the spread of criminal subcultures are certainly related to crime (Tappan 1960; Cohen 1955, 1966; Caldwell 1965; Report on Violence 1970; De Cecco and Richards 1975).

In figure 10-1, subsections 7-9 are designated as *basic environmental factors* because (1) they represent a disruption of the crucial forces of the normative constellation in contemporary society that control illegal and antisocial behavior, and (2) these variables seem most influential in predisposing and leading individuals toward those types of behavior.

10. *Opportunities.* Undoubtedly, "criminal behavior is partially a function of opportunities to commit specific classes of crimes, such as embezzlement, bank robbery, or illicit sexual intercourse" (Sutherland in Cohen et al. 1956, p. 31). This factor seems particularly relevant to those crimes considered "inside jobs," and to the vast amount of offenses that are the so-called "white-collar crimes," or perhaps more adequately, "occupational crimes," (Newman 1958; Sutherland 1961; Task Force Report 1967, pp. 102-115; Geis 1968; Geis and Meier 1977).

11. *Low social morality.* Such factors as racketeering (Kefauver 1951), organized crime (Kennedy 1960; Report by the President's Commission 1967, pp. 187-210; Cressey 1969; Gage 1972), movies (Blumer and Hauser 1933; Wolfenstein and Leites 1950; Bandura and Walters 1963), television (Eron 1963), and mass media in general (Pittman 1958; Report of the Task Force on the Media 1969; Eysenck and Nias 1978), can foster and spread crime in society.

12. *Cultural conflict.* Though the immigrant himself generally retains sufficient integrity of personality and a standard of values and attitudes more conducive to law-abiding behavior, his children are often harassed by the cultural and personal conflicts of the so-called marginal man, torn between the mores of the adopted country and those of the parental culture. Many studies have shown the relative importance of these factors; see, for example, Sellin (1938), Wirth (1931), Elliot and Merrill (1950), and Weinberg (1958). The problems of internal migration, of the high rate of mobility, of the constant change from rural to urban communities, may also be pertinent here (see Kaplan 1958; Barnes and Teeters 1959; among many others).

In figure 10-1 subsections 10-12 are grouped under the heading: *contributing environmental factors.* These factors are probably better considered contributors to crime rather than producers of criminal behavior. They seem to operate by loosening inhibitions and controls, by affording opportunities, occasions, and excuses, but not so much by creating inside the person those tendencies and predispositions that lead to illegal and antisocial behavior.

A Biopsychosocial Theory of Delinquency and Crime

Thus far, practically all the heterogeneous lines of empirical knowledge accumulated by research and investigation have been presented. The task at hand is to organize them within the framework of a general theory that will do justice to all of them. The following theoretical formulation is offered:

Criminal and delinquent behaviors are the result of a negative imbalance within the individual in the interaction between (a) the expressive forces of his psychological, biological, and creative impulses, and (b) the normative forces of familial, religious, and sociocultural factors.

Thus presented, the theory is clearly a simplified version of the true state of affairs, and should be viewed merely as a generalized conceptualization of this extremely complex problem. Figure 10-1 is an attempt to reproduce this simplified version in a slightly more developed form, as well as an effort to organize and unify the most relevant factors associated with criminality. The clusters of factors 1-6 have to do mostly (but not exclusively) with the expressive forces of the individual, whereas clusters 7-12 refer mostly to the normative forces within the family and society. I hasten to add here, nevertheless, that a theory should simplify the facts, for if a theory were as complex as the actual facts, it would have no real value (Cohen 1944).

Parsons (1954) and Merton (1957) have referred to two types of theory: (a) *generic* in character and (b) of the *middle range.* Generic theories establish a broad framework within which a variety of different classes of special events may be described. Theories of the middle range attempt to describe an intermediate area of related problems. An example of this second type of theoretical conceptualization might be a hypothesis that formulates an adequate explanation for many different forms of deviant behavior, of which criminality and delinquency would constitute only a part. Such a theory might provide a working model for comprehending a large, but limited, range of related areas of behavior. It is within this realm that the theory I present has been developed. Though it does not attempt to be all-inclusive, it provides an explanation for many different forms of criminal behavior.

The core of the theory lies in the negative imbalance, in the interaction within the person between the either driving or restraining forces of his personality and the either controlling or provoking forces of his environment. Not differently, but taking less account of the environment, Freud thought of psychoanalysis and personality as a dynamic conception that reduced mental life to the interplay of reciprocally urging and checking forces, or as he puts it, "an interplay between forces that favour or inhibit one another" (1910, p. 213; Hall and Lindzey 1978, p. 47). It can be said that under equal environmental pressure toward criminality, in a particular area, an individual high in mesomorphy (or expressive forces) and/or low in inner controls will be more likely to adopt aggressive behavior patterns than a person low in mesomorphy and/or high in inner controls; and vice versa, under equal biological and inner psychological pressure a person who encounters more opportunities or lives in a high-delinquency area will more likely adopt criminal patterns than a person raised in a cohesive family or living in a well-organized society. Similar generalizations might be offered with regard to other factors of figure 10-1. For example, it seems that clusters 5 and 6 (frustrations and pressures) had something to do with the riots and vandalism of the sixties and seventies. It is obvious also that opportunities (cluster 10) are very relevant for the large category of offenses that go under the labels of "white-collar crime" and "occupational crime." What seems basic is that in many instances both processes—biopsychological and sociological—have to be taken into account, and that it is the complex interaction and relationship between the two that should be particularly considered and stressed. The phrase "creative characteristics" has been included in the general formula to account for the new types of offenses that are constantly being introduced by many criminals.

Similar Biological Theories

Many years ago, B. Glueck (1918), a psychiatric consultant at Sing Sing Prison, asserted: "The criminal act in every instance is the resultant of interaction between a particularly constituted personality and a particular environment'' (p. 121). I find this principle vague because it applies not only to criminal but to all human acts. However, it clearly recognizes both sets of factors—the biopsychological and the sociocultural—in criminal behavior.

More recently, Beeley (1954) categorized the range and variety of empirical factors related to criminality into (a) those that enfeeble self-control and (b) those that enfeeble social control. Reiss (1951) identified law violation as a product of ineffectual controls in the person and in society. The psychiatrist D. Abrahamsen (1960) formulated his law: The Criminal Act (C) is the sum of a person's criminalistic Tendencies (T) plus his total Situation (S) divided by the amount of his Resistance (R).

In contrast to Sutherland's view, Abrahamsen's formula includes sociological as well as psychological factors, and attempts to combine the wide variety of etiological factors operating in crime into some sort of an organized framework. However, the ambiguity of the formula is implied by Abrahamsen himself when he states that the same formula may describe any form of human behavior. Nevertheless, the following statement by him merits notice: "A criminal act can take place *only* if the person's *resistance* is insufficient to withstand the pressures of his criminalistic tendencies and the situation" (pp. 57–58; italics added). Reckless' containment theory (1967), and the control theory proposed by Hirschi (1969) stating that delinquent acts result when an individual's bond to society is weak or broken, are closer to our point of view, as well as similar control theories (Hirschi 1969, pp. 18ff.). Reckless asserts: "Containment theory—internal and external containments, probably with *greater weight* assigned to the *internal*—is proposed as the best general theory to explain the largest amount of delinquency and crime" (1961, p. 356; italics added). The theory has the advantage (recognized by Haskell and Yablonsky 1978, p. 582) "of merging the psychological and the sociological viewpoints of crime causation." Reckless' is a formulation with which both psychologists and sociologists perhaps can agree.

Biopsychological Emphasis

As indicated, the theoretical explanation proposed in these pages does not differ substantially from some of these conceptualizations. The main differences lie in our efforts to gather the many heterogeneous lines of empirical investigations, to analyze more deeply the operational factors and dynamics of criminal behavior, as well as in the application of the theory to various kinds of criminality, and chiefly, in the fact that we put more stress on the biological and psychological than on the sociological factors. More concretely, the biopsychosocial formulation differs from the views considered above in the emphasis it attributes to several aspects.

1. It postulates a negative imbalance (weak inner controls versus strong and powerful drives), thereby further restricting our attention to destructive and nonconforming behavior. Admittedly, this aspect is still broad and vague.

2. The main emphasis is placed on the individual rather than on society or the individual's environment. Both clusters of factors, biopsychological and environmental, are relevant, but chiefly as they affect the individual. The imbalance that has been stressed takes place within the individual. A society, or some of its areas, may show extreme cultural or social disorganization, but if its normative forces, though they may be minute, were internalized by the individual through his family and other particular influences, his behavior might not necessarily be criminal.

3. It follows that lack of controls in society, or lack of external elements for enforcing conforming behavior, will not always per se be an etiological or decisive variable, but rather a contributing or precipitating factor to illegal behavior. Family disruption, or living in a high-delinquency area, may produce poor internalization of normative controls, and therefore may lead to antisocial behavior, but it seems that the critical factor does lie in the poor internalizaton of the normative forces of family and society rather than in the absence of external controls in the family or particular area of society. The emphasis, then, is on low personal self-control, not on low environmental control. The person and the psychological influence of his family are considered more basic and important; the community and society are looked upon as relevant but more indirect and distant factors.

4. Many sociological theories tend to emphasize the social process of transmission of criminalistic patterns, but say little or nothing about the individual process of reception of those criminalistic patterns. They are more concerned with the kinds and amounts of delinquency and crime in a society or among different sections of that society than with the processes involved in the acquisition of delinquent and criminal behavior patterns by specific individuals (Wolfgang and Ferracuti 1967, p. 28). In the following pages an effort will be made to emphasize the individual processes of reception.

5. Another characteristic of this theoretical formulation is that I will be offering some discrete operational units that may enable me to discern more clearly which personality factors—biological, temperamental, and motivational— either in themselves or, more probably, in conjunction with environmental factors (lack of parental support and supervision, association with delinquents, frustrations, opportunities) will predispose or lead to antisocial and illegal behavior.

Martin and Fitzpatrick (1965) have noted that most delinquency theories emphasize one or several of the following levels: (I) society, (II) operating milieu, (III) the family, and (IV) the individual. In our conceptualization all these levels do play some part in the genesis of delinquent behavior, but the main emphasis is on the individual, who, in turn, is varyingly affected by family environment and less directly by the operating milieu and society. In this sense the present theory is mostly psychological. As Hirschi (1969, p. 18) said, it is the bond of the person to society that is either very weak or broken.

Reiss (1951) argued that the major sources of personal control lay (a) in the social controls of the community and its institutions, (b) in the primary groups, and (c) in the personal controls of the individual through a strong ego and superego. He found that the best predictions of recidivism could be made from the last source, personal self-control. This conclusion is similar to the assumption made by Gold (1963), namely, that "the causes of delinquency, as with any other behavior, must be understood in terms of the forces operating *psychologically* to provoke or control it" (p. 35). As we have seen, Reckless (1967) also assigned greater weight to internal containments.

An effort has been made to organize by means of a simple formula the many factors found to be relevant to delinquency and crime. It is based on the imbalance of biopsychogenic factors and sociocultural factors. The formula may be described in terms of general, Freudian and learning theories as follows:

$$\text{General:} \quad \frac{\text{high expressive forces}}{\text{low normative forces}}$$

$$\text{Freudian:} \quad \frac{\text{strong, aggressive id}}{\text{very defective superego}}$$

$$\text{Learning:} \quad \frac{\text{low conditionability}}{\text{very deficient conditioning}}$$

In all three formulas what may be called the numerators are mostly biopsychogenic and the denominator mostly sociocultural. It is the disproportion and interplay between the two aspects of each equation that seems to account for delinquency and crime. It should be added that the basic general principle may have validity both inside and outside the person. In the case of delinquent behavior, there are undoubtedly excitatory or urging forces in the environment: association with criminals, low social morality, frustrations, pressures, and the like. There are also checking or inhibitory forces in the environment: social control, law enforcement, correctional agencies, and anticriminal patterns. It is the interplay of this excitation-inhibition element in the environment together with the excitation-inhibition imbalance within the personality that accounts for antisocial behavior: propensity in the person plus incentives in the environment.

Etiology of Delinquency and Crime

Most sociological theories concerning crime fail to provide an adequate answer to this recurring question: Why, under similar environmental circumstances, do some individuals become delinquents and others do not? The answer to this question seems to be the essence of the problem of the study of delinquency.

Numerous studies (Jenkins 1966; Quay, 1965; Finney, 1966) have repeatedly isolated several clusters of traits, or syndromes, associated with delinquency. The three most common clusters have been labeled: the *unsocialized, psychopathic* delinquent (low in guilt), the *neurotic* or *deviant* delinquent (high in guilt), and the *subcultural* or *socialized* delinquent (high in socialization). At this time I will focus on the first group, which comprises the most typical offender: the *aggressive,* hardened, recidivist, psychopathic delinquent.

It seems convenient to distinguish between delinquent or criminal behavior, and deviant or unconventional behavior. The category of *criminal* acts should probably be limited to those acts that are either harmful to another person or persons, or to their property. If there is no harm or damage involved, as in the case of homosexual actions performed in private between consenting adults, particular cases of excessive drinking (with no driving involved), instances of drug use, prostitution, and the like, it appears better to classify those actions as "variant," "unconventional," or even "deviant" and "immoral," but not criminal and delinquent.

It is my opinion—and I consider this the core of the etiology of delinquency—that all individuals possessing a strong id in the area of aggression and a poorly developed superego are, in fact, *predelinquents.* Or, to use a slightly different terminology, predelinquents are those individuals high in expressive forces and very low in restraining self-controls. Such persons may or may not in fact become delinquents and criminals, depending on various other circumstances such as opportunity, incentive, type of companionships, different pressures or frustrations. I can say definitely, however, that these individuals are *more likely* to commit delinquent acts than individuals without such personality characteristics.

When there is no attachment to parents and consequently to others and to society and its laws, moral restraints if they exist will often be neutralized and the laws broken. In the words of Abrahamsen (1960), "A criminal act can take place only if the person's resistance is insufficient to withstand the pressures of his criminalistic tendencies and the situation" (pp. 57–58). A defective superego constitutes insufficient resistance. Given the situation, particular circumstances and external pressures that may be considered a *condition* for crime rather than a cause (a condition that, for instance, will be found more often in high-delinquency areas), the following formula for delinquent and criminal behavior is obtained:

$$\text{delinquent behavior} = \frac{\text{strong, aggressive id}}{\text{defective superego}} + \text{circumstances and incentives}$$

Besides being a general conceptualization, I believe that this explanation offers discrete operational units that may enable us to discern which *personality* factors, biological (such as mesomorphy), psychological, and motivational (such as impulsiveness, low conditionability, defective ego and superego, high need for achievement and power), in conjunction with specific *environmental* factors (disrupted family, lack of parental support and control, deficient religious training, presence of opportunities, frustrations, and pressures) predispose to violent and antisocial behavior. It is also clear that not all individuals from the same family, area, or community will possess or encounter all these variables to the same degree, and therefore some will, and others will not, become delinquents.

It is both difficult and dangerous to designate the causes of delinquency because research has discovered only *associations* or relationships between specific variables and delinquent behavior. Nevertheless, I would advance that the factors or conditions that appear to be closest to the main root of delinquency are: *Family disruption, particularly when the children are mesomorphic.* I interpret family disruption as either physical or psychological disruption. Since family disruption is increasingly affecting all social classes, this could explain why crime is also increasing at all social levels.

By creating a defective superego and no real bonds to society, *family disruption* lowers the individual's resistance to withstand the pressures of his aggressive tendencies in the presence of frustrations, incentives, and opportunities, and increases individual differential receptivity toward criminalistic patterns. On the other hand, by being intimately connected with potential aggressivity, *mesomorphy* increases the person's chances of reacting aggressively and antisocially toward those same incentives and opportunities. It is because of high aggressivity and differential receptivity within the personality of some individuals that these people will differentially associate, will take advantage of situations and opportunities, will react with violence to varying pressures and frustrations, and more often than not, will select antisocial types of behavior from among the various courses of action open to them.

As for delinquent behavior in general (not only violent and aggressive behavior), it may be stated that a deficient ego and superego (and the family disruption these factors imply) are probably the necessary and sufficient condition. From a theoretical standpoint it is the necessary condition because it appears reasonable that in order to fail persistently to conform to the rules of society one must have failed to internalize its rules or norms. It is also necessary from an experimental standpoint because most studies, including my own, have shown that the superego, as well as the ego, of delinquents and criminals is defective.

This defective superego is also a sufficent condition to explain illegal actions because unethical and criminalistic tendencies (not necessarily aggressive) are present in everybody, to a greater or lesser extent, and situations and pressures leading to delinquent behavior sooner or later will also be encountered by everyone.

As the reader will have noticed, in explaining criminal behavior I have avoided the simplistic "evil-causes-evil" assumption. Delinquents possess many "good" qualities: a mesomorphic physique, great physical energy, somatotonic characteristics, high need for achievement and power. It is their great deficiency in another area of personality that seems to change their potential for constructive entrepreneurship into one that is destructive for society and its citizens.

Different Types of Illegal Behavior

Up to this point I have dwelt mainly on the typical delinquent, the aggressive, persistent, hardened offender. Criminal behavior, however, is a very complex pattern that involves multiple processes and comprises many other types of offenders. Frequently delinquency in adolescents does not lead to adult criminality (Glueck and Glueck 1968); it is intricately fused with all manner of acquired beliefs, past experiences, drives, tastes, habits, impulses and inhibitions. To say, for example, that a certain boy or girl is delinquent is by no means to characterize his or her motivations and dispositions. There are myriad forms of delinquency, ultimately as many forms as there are individuals whose impulses and tendencies have been channeled toward illegal behavior.

Not all persons are strongly predisposed toward criminal behavior; and when they are, such a predisposition must be aroused before it becomes active. To a degree there is always either an internal or an external sensory cue that instigates the tension. Moreover, some (but not all) motivational dispositions are self-active. In many cases it is not the sight of a car that impels a delinquent to theft, nor a bad companion who motivates an offender to assault a third person in order to show off in the presence of the companion. Not infrequently the response seems to spring from deep-seated trends, which in another individual may prompt only admiration of the car's features or an innocent show of strength. Such deep-seated trends and dispositions are not merely ways of reacting to environment, but are also ways of meeting it. It is impossible, therefore, to consider selected dispositions and characteristics as units wholly apart from the total personality. Nevertheless, it may be fruitful to offer a number of theoretical considerations regarding a few of the main types of criminal behavior, but without any attempt to cover all the different categories.

Some crimes involve violence whereas others do not. Among criminals of the *violent* type we have considered mainly the persistent, recidivist, and somewhat psychopathic offenders. There are also the occasional and, so to speak, accidental offenders, those who commit violent crimes in the heat of passion, on the spur of the moment, or as a last resort after having tried all legal means. Among "criminals"' of the nonviolent type I may include (a) some sexual offenders whose crimes do not involve aggression; and (b) occupational offenders such as embezzlers, forgers, and tax evaders, as well as some professional criminals such as burglars, and a few instances of organized crimes. These groups, (a) and (b), may include both persistent and occasional offenders. The explanation that has been proposed for the more psychopathic and violent offenders does not apply to all these other types.

It has already been shown that the fraction, strong aggressive id/poorly developed superego, leads more often to violent criminal behavior. Other types or varieties of criminal behavior will now be considered. I begin with the occasional aggressive offenders, that is, those persons who commit crimes only in moments of great passion or frustration, or when forced to commit a crime they would prefer to avoid. These offenders could show the combination of the persistent offender, but it seems that they might better be represented by the fraction strong aggressive id/normally developed superego. Here the controls within are firm and the inhibitions against criminal behavior are high, but special circumstances and situations involving force, threat, or passion may weaken this strength momentarily, upset the balance of the mechanisms, and lead to crime.

The occasional criminal in the course of his occupation ("white-collar" offenders) may have a strong superego. Such persons are usually respected, may come from highly regarded families, often exhibit internalized controls, and frequently are found in positions of financial trust. Since normally they do not exhibit aggressiveness or violence the formula would be: weak aggressive id/strong superego. Due to opportunities and special pressures or circumstances, the resistance of the strong superego is lessened and the crime committed. However, as mentioned, I have in mind the occasional "white-collar" criminal. Obviously, if his criminality is persistent, if he is a professional or a recidivist, the lack of controls and inhibitions will be more permanent, the superego weaker, and the formula in this case is: weak aggressive id/weak superego.

Similar reasoning may be applied to the nonaggressive sexual offenders. If the offenses are persistent and continuous, the superego will be weak.

Table 10–1
Personality Characteristics of Various Types of Criminal Behavior

Condition:	Persistent, violent offenders	Persistent, sexual offenders
Personality:	$\dfrac{\text{Strong, aggressive id (mesomorphs)}}{\text{Defective superego}}$	$\dfrac{\text{Strong, sexual id (ectomorphs)}}{\text{Defective superego}}$
Condition:	Persistent "white-collar" offenders	Occasional offenders
Personality:	$\dfrac{\text{Weak, aggressive id (nonmesomorphs)}}{\text{Defective superego}}$	$\dfrac{\text{Sexual, aggressive, or weak id}}{\text{Less defective superego}}$

Since the offenses are not aggressive, the id will also be weak in the area of aggression, although not in the area of sexuality. Such an id might be called a strong sexual id. Even the physical characteristics of these two types of criminals: the nonaggressive sexual on the one hand, and the aggressive and violent on the other, will most likely differ if our findings with regard to sexuality and constitution (Cortés and Gatti 1972, pp. 96–100) are proved valid. As reported there, a positive correlation exists between ectomorphy and manifest sexuality. If further studies confirm Sheldon's findings, non-aggressive sexual offenders would be more ectomorphic than aggressive offenders.

Table 10–1 summarizes the preceding hypothetical considerations.

A defective superego, lack of inner controls, is assumed to be a necessary condition or predisposition for all types of persistent criminal behavior. Whether or not the crime will tend to be violent, nonviolent, or sexual, appears to be a result of the biopsychological characteristics of the offender, as well as of varying environmental factors. A study of the different types of crimes committed by persons with differing constitutions, similar to the studies already made regarding typologies and personality characteristics (Quay 1965; Lemert 1969; Gibbons 1977), would prove interesting and rewarding.

Further Hypotheses

The usefulness of a theory depends in part on its fertility; that is, on the number, relevance, and validity of hypotheses and further areas of investigation that can be deduced from it. A great number of hypotheses may be derived from the proposed theory, many of which have already been mentioned. In this section I will explain some of them in more detail and will suggest others; all are testable and subject to empirical research.

Delinquency and Blacks

The official indexes of crime appear to have established the fact that at the present time the number of offenses committed by the American black population is extremely high. Blacks comprise approximately 12 percent of the total population, but in the past two decades they have accounted for about 30 percent of all arrests. The picture is gloomier with regard to the more serious and violent offenses. "Reported national urban arrests are much higher for Negroes than for whites in all four major violent crime categories, ranging from ten to eleven times higher for assault and rape to sixteen or seventeen times higher for robbery and homicide" (Report by the National Commission on Violence 1970, p. 20). The latest statistics for arrests on violent crimes (*Crime in the U.S.*, 1977, October 1978) published by the FBI state that, nationally, of persons arrested for these particular crimes, 49.7 percent were black.

The Philadelphia study by Wolfgang (1958) shows the disproportionate presence of blacks in the total picture of criminal homicides. The black population of Philadelphia was 18 percent, yet 75 percent of the homicide offenders were black. Even black females accounted for more homicides than white males. In a subsequent study by Wolfgang, Figlio, and Sellin (1972) on 10,000 boys born in Philadelphia in 1945, about 50 percent of the 3,000 nonwhites had at least one police contact by age 18, compared with 20 percent of the 7,000 whites. The differences were more pronounced for the major *violent* offenders. For example, 86 percent of the rapes, 82 percent of the aggravated assaults, and 90 percent of the robberies were committed by blacks. It should be mentioned that in homicides, aggravated assaults, and rapes (but not in robberies) the majority of the victims were also blacks. This aspect is also true nationally.

The objection could be advanced that these high rates are the result of direct discrimination: the police patrol more heavily in black areas, think that blacks are unusually likely to commit criminal acts, and so on. One cannot deny that it may be true in many areas. But at present this explanation does not appear to be the only one, nor probably the most basic reason for the difference. Bloch and Geis (1962) wrote: "In parts of the Deep South, there is considerable evidence that except for serious violent crimes against whites, the police and the courts are generally lenient in the prosecution of Negroes, especially in rural areas (p. 180)."

Attempts to explain this high rate by emphasizing that blacks are a minority group do not seem to be acceptable either. For serious offenses the rate for blacks is at least twice that of Indians, and the Indian is at the bottom of the socioeconomic scale (Reid 1976, pp. 60, 62). Rates are also very low for Chinese, Japanese, and Jews. Aside from a critique of official arrest statistics (see Geis 1965; Hinderlang 1979), there is no real evidence to deny convincingly that there is greater than average black involvement in crime, very particularly in violent and assaultive crime.

If the theory of delinquent and criminal behavior that has been proposed is accurate, it would follow that blacks should be higher in mesomorphy (because of the high rate of violent crime) and in family disruption (because of greater involvement in crime). This hypothesis is worthy of further research, but the evidence already at hand points in that direction. With regard to mesomorphy, Sheldon (1940) photographed and somatotyped about 400 northern blacks. He wrote: "Among Negroes it is the extreme second component (mesomorphy) that is *most prominent* . . . "(pp. 220–221; italics added). The same conclusion can be inferred from the *very high* proportion, in relative terms, of black athletes in such sports as boxing, basketball, football, track and field, and baseball (see Davis, 1966, pp. 774–825).

With regard to higher physical and psychological family disruption among blacks, the evidence seems to be stronger. One of the most significant studies is undoubtedly the Moynihan Report (1965). He clearly demonstrated the existence of much greater family disruption among lower-class blacks than among whites of the same class. It is also known that they have a significantly higher illegitimacy rate. The Moynihan Report gave the estimated proportion of illegitimate births in central Harlem as 43 percent. In the words of Alsop (1969): ". . . the latest estimate is 56 percent. About 80 percent of the first-born children in Harlem are illegitimate" (p. 92). A national magazine (*Time,* April 6, 1970) reported that the latest available illegitimacy rate for the nation was 4.9 percent for all live white births as against 29.4 percent for nonwhite births. The same magazine said that U.S. census figures estimated that 29 percent of black households were fatherless as against 9 percent fatherless white households. More recent statistics have not altered this picture.

As for psychological disruption, the Report by the National Commission on Violence (1970, p. 27) commented that a great many inner-city families are large and involve conflict-ridden marital relationships, that parents communicate to their offspring little sense of permanence and few precepts essential to orderly, peaceful life. Child-rearing problems are exacerbated where the father is periodically or frequently absent, intoxicated, or replaced by another man, or where children have arrived too early and the parents are too immature to put their child's needs above their personal pleasure.

The above data are pertinent not only because they tend to confirm the hypothesis advanced, but also because the commission recognizes that blacks in the cities have improved in educational level (61 percent completed high school in 1968 against 43 percent in 1960), that "the rates of unemployment dropped significantly between 1960 and 1968," and that the number of persons living below the poverty level in the cities had declined in that same period from "11.3 million to 8.3 million" (p. 33). Obviously, much more remains to be accomplished in all these areas. However, the commission fails to draw the conclusion that the significant increase in crime may be connected with the significant increase in family disruption. They think, rather, of: ". . . disappointments of minorities in the revolution of rising expectations, the weakening of law enforcement, and the loss of institutional legitimacy in the view of many" (p. 37).

Thus, the hypothesis of postulating higher mesomorphy and greater family disruption in blacks in order to account for the higher rate of crime, and, in particular, of violent crime, appears reasonable and supported by the available evidence. There is little need to emphasize the obvious; that it is *not* race, per se, nor even the mesomorphic component, which when properly channeled can lead to very constructive and enterprising behavior, that accounts for the high rates of crime. It is, rather, higher disorganization among black families, the increased physical and psychological disruption *within* the home. From this point of view the impetus toward integration, toward decreasing cultural and economic deprivations, toward better housing and removal of slums, and so on are excellent and should be continued and increased, but not as a means or as direct influences for reducing crime. For this purpose the all-important goal is better *intrafamily* organization, a more stable and integrated family life, in short, much less physical and psychological disruption within the home. These are improvements that will go a long way toward counteracting the powerful external influences that pull young men and women toward delinquency. One final aspect should be mentioned. Blacks are still in the minority in two categories of crime, both less reported and detected, but very broad and extremely damaging to the nation: "white-collar crime" (see Geis 1968) and "organized crime" (Cressey 1969; Bloch and Geis 1970, p. 198; Gage 1972).

Delinquency in Women

Women are not as mesomorphic as men (Sheldon 1954; Parnell 1958). Consequently, by the proposed theory, the proportion of aggressive and violent crimes committed by female offenders should not be as high as that of male offenders. This is an established fact confirmed by the official crime reports. In spite of recent increases in female criminality, in 1977 the proportion of male arrests for violent crimes was 89.4 percent as compared with 10.6 percent for females (see W. Webster 1978). The disproportion concerning sex ratios in crime is also true throughout the world. "The crime rate for men is greatly in excess of the rate for women—in all nations, all communities within a nation, all age groups, all periods of history for which organized statistics are available, and for all types of crime except those peculiar to women, such as infanticide and abortion" (Sutherland and Cressey 1974), p. 126. This fact needs no further proof (for a different view see Pollak 1950). However, other additional hypotheses seem worthy of investigation.

1. Mesomorphy is related to aggressive criminality. Therefore, those females convicted of crimes of violence should be *more* mesomorphic than a comparable group of women. Epps and Parnell (1952) studied the physique of 177 convicted female offenders. They were found to possess a strong tendency toward mesomorphy. They were also higher in that component than a comparable group of undergraduate females. Nevertheless, the method employed left much to be desired: only height, weight, and fat measurements were taken.

2. A second hypothesis may be more relevant. Women are not as high in expressive forces (mesomorphic component) as men, and therefore the normative forces in female delinquents should be *lower* than in male delinquents; in other words, female offenders should come from disrupted homes *more* often than male offenders. A great number of studies appear to support this hypothesis. A report by the California Youth Authority (1968) makes it clear that compared with delinquent boys, delinquent girls come from more disorganized backgrounds. Webb (1943) studied 40 female delinquents in Seattle. Nearly all of these girls came from broken and disordered homes; in addition, 21 of the girls had alcoholic, criminalistic, or severely neurotic parents. Kaufman et al. (1959) and Konopka (1966) emphasize that female delinquency is a tension-management response to conditions of parent-child disharmony. Monahan (1957) made one of the most comprehensive studies involving 44,448 delinquents of both sexes, that is, all the cases referred to the Philadelphia juvenile court during the period from 1949 to 1954, whether or not the cases were heard by the court. At that time according to the census, 7 percent of the white and 33 percent of the non-white Philadelphia children under 18 years of age were not in husband-wife families. His data indicated very clearly that the breaking of the home was differentially related to delinquency, affecting "girls most damagingly." Of the white female recidivists, 68.6 percent came from broken homes versus 41.4 percent of the males; of the black female recidivists, 80.2 percent came from broken homes versus 62.2 percent of the males. Weinberg (1958) pointed out: "The broken home tends to have a more perceptive influence upon female than upon male delinquents, and significantly more delinquent girls than delinquent boys come from broken homes" (p. 126). Gibbons (1970) after reviewing a great number of studies, summarized: "'Under the roof culture' in form of family tension of one kind or another appears to be a major factor in female delinquency" (p. 189).

No one would deny that further research to test this type of hypothesis is relevant today especially as female delinquency and arrests are increasing at a faster pace than male arrests. In the period 1968–1977 male arrests went up by 13.4 percent while female arrests have increased by 57.6 percent (see Webster 1978, p. 175). Fortunately, the number of studies of women and crime appear to be increasing (R. Simon 1975; F. Adler 1975; Gibbons, 1977, pp. 439–463).

The Psychopathic Personality

Today a growing number of psychologists and sociologists admit the existence of a very dangerous type of criminal, the psychopath (or sociopath), and recent investigations have clarified the nature of psychopathy (McCord and McCord 1964; Craft 1966; Arieti 1967; Cleckley 1976; Hare 1970; Hare and Schalling 1978; Reid 1978). The psychopath may be defined as a *loveless, guiltless, highly aggressive* person. For a different view still prevalent among several sociologists see Wallinga (1956) and Sutherland and Cressey (1974, pp. 159–161). The theory that has been presented applies particularly to this category of persons: very high impulsivity and a total lack of inner controls.

The true, simple, or primary psychopath may well be located in the most aggressive end of the general dimension of criminality. The book *Killer* by Gaddis and Long (1970) contains the diary of an extreme psychopath, Carl Panzram. It is clear from the book and from many other investigations that psychopaths differ, at least in degree, from other criminals. According to McCord and McCord (1964): "The total pattern of the psychopath's personality differentiates him from the normal criminal. His aggression is more intense, his impulsivity more pronounced, his emotional relations more shallow. His guiltlessness, however, is the critical distinguishing trait" (p. 51). The psychopath also differs from the typical neurotic in that the latter feels intense inner anxiety, represses his hostility, and is often oppressed by guilt, but usually can maintain bonds of love. The psychopath does not. He differs from the psychotic in that the psychopath attacks reality while the psychotic generally withdraws from it and often experiences hallucinations and feelings of intense guilt or at least anxiety. Many psychopaths exhibit varying signs of brain malfunction (Hare and Schalling 1978), and the possibility of physiological and neurological disturbances should be kept in mind.

The following hypotheses concerning this category of criminal may now be stated: (1) not only their superego but also their ego would be very defective; (2) they would be high in mesomorphy; (3) they would be lower than other criminals in the endomorphic component; (4) their family environment would be characterized by a high degree of rejection and disruption. Here are the reasons for suggesting these hypotheses.

1. *Defective ego.* Freud very often considered the superego as a synonym for "conscience." In Freud's theory the superego is the heir of the Oedipal complex and becomes a distinctive province or agency of personal-

ity only after the complex is resolved. However, this resolution of the complex is brought about by the boy's fears, his guilt feelings concerning his incestuous desires, and his castration anxiety. These feelings help to bring about the identification of the boy with his father—a factor crucial for the resolution of the complex. It appears, therefore, that the superego is already at work (by inducing repression of impulses and fostering identification) before the Oedipus complex is solved. In my view it seems more consistent to conceive of conscience and moral controls both as an inborn potential and as the internalization of parental norms and prohibitions.

I should distinguish between what Maslow (1971, p. 338) calls "extrinsic conscience" and the "intrinsic conscience." The former is part of what Freud refers to as the superego and to a greater degree will be developed first by the influence of the parents and later by the influence of peers, culture, and society. The "intrinsic conscience," on the other hand, is "ultimately biologically rooted" (Maslow 1971, p. 338), inherent in human nature—what Jung terms (Evans 1976, p. 281) "built-up superego, . . . the eternal truth in man himself, because he checks himself." It is developed, as are other potentials in man, by proper reasoning, by social or school influences, and by religious or moral training. In some aspects, both types of conscience may coincide. However, the intrinsic conscience would not be a part of the superego proper but of the ego, as this province of personality is understood by White (1963), Erikson (1963), Hartmann (1964), and many others (see Hall and Lindzey 1978, pp. 75–112). In his book, *Conscience and Guilt,* Knight (1970) reaches a similar conclusion.

The absence of conscience and total lack of moral and guilt feelings that so distinguishes the psychopath would lead us to think that he has a very defective ego and superego—that is, a malfunctioning of both the intrinsic and the extrinsic consciences. It is very probable that the ego deficiency can be related to the autonomic and cortical correlates of psychopathy, and this hypothesis seems worthy of further investigation. Ability to delay gratification is supposed to belong to the ego, and in Mischel's review (1976) of much relevant research, psychopaths are deficient in this characteristic.

2. *Mesomorphy.* In order to account for the extreme aggressiveness and impulsivity of the psychopath, I have assumed that he must be high in the mesomorphic component. Recent research on psychopathy is relevant here. Among the many important findings concerning psychopaths the following should be mentioned:

1.   reduced cortical excitability; that is, low level of cortical arousal (Petrie 1967; Lykken 1968), and therefore great need for stimulation (Hare 1968);
2.   infantile slow-wave activity (Kiloh and Osselton 1966), and the possibility of *delayed* cortical maturation (Elliot 1978);

3. localized electroencephalographic abnormalities, possibly connected with malfunction of certain brain mechanisms (Hare 1970);
4. very low conditionability, particularly of fear responses (Schachter and Latané 1964; Schmauk 1968);
5. low levels on several indices of autonomic activity and responsivity (Hare 1968, 1970; Hare and Schalling 1978).

I have the strong suspicion that most of these findings are characteristics associated with the mesomorphic component and not exclusively with psychopathy. In the majority of the above-mentioned studies, psychopaths have been contrasted with nonpsychopaths and noncriminals who were very likely lower in the mesomorphic component. There is an urgent need to contrast psychopaths with *noncriminal mesomorphs*. If the same psychophysiological and neuropsychological findings apply to the latter, much of the research concerning mesomorphy reported in Cortés and Gatti (1972) would acquire a deeper meaning.

3. *Low endomorphy.* I have hypothesized lower endomorphy in the simple psychopath than in other delinquents and criminals. The reason for this hypothesis is the inability of the psychopath to form bonds of affection with other human beings and his total lack of empathy and compassion. I have shown (Cortés and Gatti, 1972, pp. 43-74) that endomorphs often described themselves as affectionate, kind, sociable, sympathetic, and that they were higher in social values. Sheldon's delinquents with the highest criminal record score and our 20 criminals (Cortés and Gatti 1972, pp. 17, 30) are lower in endomorphy than all other delinquents. It is probable that many of them were psychopaths, and the data would give some support to this minor hypothesis.

4. *Rejection at home.* The hypothesis that psychopaths have suffered severe rejection at home is in agreement with the conclusion of many other investigators. McCord and McCord (1964) have stated that "brain damage alone does not result in the distinctive characteristics of the psychopath: guiltlessness and lovelessness (pp. 84-85)," and they suggest three causal patterns: (a) severe rejection, by itself, can cause psychopathy; (b) mild rejection at home, in combination with damage to the brain area (possibly the hypothalamus), can also cause psychopathy; and (c) a third pattern may be mild rejection in the absence of neural disorder, plus certain other influences such as a psychopathic parental model, punitive discipline, and absence of supervision.

Occupational Crime

The broad category of *white-collar crime* (I believe a much better term would be *occupational crime*) may be defined as comprising those offenses committed by persons in government, business, and the professions, in the course of their occupational roles (Geis and Meier 1977).

As previously indicated, the crimes under this broad heading involve little or no violence. In accordance with the theory, I would not expect a higher than average mesomorphy among this category of criminals. The same may be true with regard to some offenses committed by those who belong to "organized crime." In these criminals in addition to the basic environmental factors of figure 10-1, variables 4, 6, and 10 would be among the most relevant. This aspect needs to be stressed in order to reassure those who feel I may have overemphasized biological and constitutional factors. However, I agree with the assertion by Morris (1965) that "the attributes of white-collar crime that constitute the most significant characteristic are shared by offenders who wear blue collars, coveralls, uniforms and even dresses" (p. 202). There is no reason to postulate a different set of characteristics for this category of criminals except for the absence of those biopsychological traits that are related to violence and aggression.

Nonmesomorphic Offenders

Although there is a greater proportion of mesomorphs among delinquents and criminals, not all of them are high in this physical component. If the theory that has been proposed is correct, one would expect the ectomorphic delinquents to be both less aggressive in their criminal acts and more affected by the disruption in their homes, and by exposure to criminal patterns and conditioning. Some of the studies by the Gluecks (1956, 1962) uncover a striking contrast between mesomorphic and ectomorphic delinquents. Mesomorphic delinquents tended to "underreact" to adverse familial conditions, and thus a smaller proportion of mesomorphs than of the other physical types were affected by their broken homes. On the other hand, ectomorphs "overreacted" to poor familial conditions. Washburn (1963) also supports the above conclusion. Trasler (1962) follows Eysenck's theory that the individual's position upon the introversion continuum is partly determined by genetic factors. Since introverts and ectomorphs are more conditionable than mesomorphs, they will be more deeply and permanently affected by exposure to illegal and criminalistic conditioning.

Disruption of Family Relationships

It is my belief that the disruption of intrafamily relationships is by far more relevant to delinquent behavior than most of the other variables in figure 10-1. Since many sociologists and a number of psychologists still insist on the almost exclusive relevance of social disorganization in many areas, low economic conditions, and differential association, such belief could easily be tested by contrasting groups of delinquents (matched with regard to other variables), from all social classes, from disrupted and nondisrupted homes, and from high- and low-delinquency areas. I would expect delinquency to be associated with the disruption of the home, no matter the social class of the parents or the delinqency of the area. Better and more controlled studies are needed but support and evidence for this inference are found in the study by Herskovitz, Levine, and Spivack (1959) on antisocial behavior from higher socioeconomic groups. They concluded that intrafamily experiences rather than socioeconomic factors were causal in the development of delinquency. The same conclusion can be drawn from the studies by Reckless and his associates (Reckless 1967) on "insulated" and "noninsulated" boys from a high-delinquency area. Support is also afforded by the many predictive studies (in the United States as well as in many other countries) that take into account, either exclusively (Glueck and Glueck 1972) or principally (Briggs, Wirt, and Johnson 1961), the intrafamilial conditions in the homes of future delinquents and nondelinquents.

Comments on Criticisms of the Theory

A number of criticisms of the theory just presented have come to my attention since its initial publication in 1972. It would seem that the theory has had little impact on the field of criminology and I am aware of only one author, writing in *American Anthropologist* (see Bolton 1976, pp. 148–149), who seems to have understood the essence of all our efforts.

Biological Factors

Unlike much of the western world, the United States seemed to tend toward the position that the study or defense of biological factors as one of the many elements in the complex factors concerning behavior, personality, and criminality is not to be considered. The trend has been decidedly panenvironmentalistic. McClearn (1969) appears quite correct when he wrote: "There often appears to be among social scientists an implicit (and sometimes explicit) feeling that any point conceded to biological factors is one point less for social and environmental factors. This attitude reflects a dichotomous view of behavioral determination that is as unwarranted as it is widespread" (p. 979). I think that we agree that all of human behavior is biosocial in nature; that is, it has *both* biopsychological and sociocultural components. All of these aspects deserve consideration and research.

However, in the field of criminology I fear we still have a long way to go. The most common denominator of the reviews I have seen is that of surprise that we even "initiated a serious attempt to re-examine biological and environmental factors in crime and delinquency." David R. Walters (1973), in reviewing Cortés and Gatti (1972) wrote: "G.A. Harrison, M.F. Ashley Montague, F. Loring Brace, and many other distinguished names have dealt the notion of body typing severe, it not fatal blows. In particular the notion that body type and personality are somehow related has been abandoned as an unproductive avenue by most researchers" (p. 161). Professor Walters seems to have read only the summary of the book, that is, the last few pages. In the text we strongly criticize both Sheldon and the Gluecks. The methods we used to measure physique, temperament, and motivation differ entirely from theirs and are truly objective. The distinguished names that Walters mentions have not done basic research in this particular field, and Ashley Montague is quoted in our book because he praises Parnell's method of somatotyping, which is the one used by us. Most importantly, Walters does not seem to be familiar with the literature that has accumulated in the intervening years (see Ryckman 1978; Engler 1979). C. Hall and G. Lindzey (an expert in the field and past president of the American Psychological Association) say: "We believe that if readers will examine carefully the numerous studies cited at the end of this chapter and the surveys by Lindzey (1967) and Rees (1968, 1973) they will come away convinced that Sheldon is eminently correct in his assertion that there is a highly signficant association between physique and personality. . . . Thus, our belief is that an overall appraisal of the many studies conducted since Sheldon began his work will lead the reader to accept the existence of a significant and interesting relation between physique and personality" (1978, p. 518). Among the numerous studies alluded to in the above quote are several on physique and criminality.

On the positive side, allow me to mention the following incident. In his book, *Delinquent Behavior* (1970), Professor of Sociology D.C. Gibbons is more impressed than I am by the fact that the Gluecks found twice as many mesomorphs among lawbreakers and comments:

> These findings [the Gluecks'] were the result of careful measurement, so that there is little question as to their accuracy. However, a sociologist would be quick to point out that a process of *social selection,* rather than biological determinism, probably explains the results. In other words, it is not unlikely that recruits to delinquent conduct are drawn from the group of more agile, physically fit boys, just as "Little League" baseball or "Pop Warner" league football players tend toward mesomorphy. Fat delinquents and fat ball players are uncommon, because social behavior involved in these cases puts fat, skinny, or sickly boys at a disadvantage. If so, the findings reflect the work of social factors, not biology. (pp. 75–76)

In response to such criticisms we (Cortes and Gatti 1972, p. 40) wrote:

> It is exasperating to find this type of objection again and again. . . For one thing, nobody nowadays speaks of biological *determinism;* but, secondly, the fact that there may be a process of social selection does not detract in the least from the relevance of many other variables. Who makes this selection? Why is the particular selection made and not an entirely different one? Is it not, as Gibbons concedes, because fat, skinny, or sickly boys are at a disadvantage? Do physique and constitutional factors play no role whatsoever in baseball, boxing, or football? It is his last sentence that shows more clearly the onesidedness of this point of view: '. . . the findings reflect the workings of social factors, not biology.' The proper and logical conclusion for the entire paragraph, after the explanation he advances, should rather be: 'the findings may reflect the workings of *both* social *and* biological factors.'

About a year later, Professor Gibbons wrote that he had read our book some time before and, among other things, read the comment we have just quoted. He agreed that many have used his statements "to discuss the work of the Gluecks and kindred souls," the latest example to come to his attention being the book, *The New Criminology,* published in England in 1973. Then he continued: "I am chagrined about this matter. I found your rebuttal to my evaluation to be quite convincing."

Macroscopic Theory

In reviews and recent textbooks some authors have quarreled with our sample (Johnson 1978, p. 140; Reid 1976, p. 149). That the samples were small cannot be contradicted although a few comments may be pertinent. We tested 100 official delinquents (70 percent institutionalized, or "incarcerated" as Reid puts it) and 30 percent who had never been institutionalized (on probation or under suspended sentence); 100 nondelinquents (no official record whatsoever); and 20 adult felons, all of them incarcerated in the Washington, D.C., area. The size of the sample to measure only the association between physique and delinquency is not so small if the proper statistical tools are used. But two aspects should be mentioned. All the 200 boys were between sixteen-and-a-half and eighteen-and-a-half years of age, an average of seventeen-and-a-half years old. Thirty percent of all delinquents had never been institutionalized.

I grant that the size of our samples was not ideal. Nevertheless, in my opinion it is a secondary matter. As another reviewer has pointed out: "Comparing samples of delinquents and nondelinquents, Cortés and Gatti argue that delinquents are higher in mesomorphy, activity, aggressivity, and need for achievement, and lower in religiousness, parental discipline and parental affection." The reviewer goes on to suggest that because of the "criticisms of previous research, this additional demonstration of an association between delinquency and body build is valuable" (Bolton 1976, pp. 148-149). This is what matters. It has been an additional demonstration, one that was needed, and it has confirmed the conclusions of earlier studies.

I am concerned about the criticism that the size of the sample is insufficient to build a macroscopic and biopsychosocial theory, for this is not plainly so. In our book and in this present chapter I have attempted to integrate the findings presented by *different* authors and investigators. I hope figure 10-1 and the entire first section of this presentation have demonstrated this fact.

Beta Weights

This criticism was formulated by Herbert C. Quay (1973), a distinguished scholar in the field of criminology. We quoted him several times in our book and it is he who noted we included practically all variables associated with delinquency. However, he goes on to argue that the relative importance of the variables is not specified nor is the issue of their interaction dealt with. He goes on to argue that the "reader is in the position of the applied psychologist who has been given a multiple regression equation without the Beta weights and who, on further examination, also finds that the measures of some of the variables have also not been provided" (Quay 1973, p. 275).

To answer some of his minor queries first, let me clarify that delinquency will occur if one or several of the factors are present. Hardly ever will all of the factors be present in a single individual. At times the factors will be addictive, at other times multiplicative; often they may be compensatory, and occasionally disjunctive. Measures of some of the variables have not been provided because the pertinent sources have been referred to, and the interested reader can consult them. But all these aspects will become clearer in the answer to his main criticism, namely that the reader has not been given the relevant Beta weights with which to interpret the multiple regression equation.

A multiple regression equation is an equation for computing a criterion variable score for *an individual* from his scores on several other variables. It is derived from the correlation of each of these variables with the criterion and from their intercorrelations. The Beta weights in a multiple correlation represent the amount by which each variable is multiplied so that each variable or predictor will produce the highest possible multiple correlation with the dependent variable; the dependent variable, in our case, would be delinquency. By using this multivariate technique we arrive at a set of optimum weightings for each of the predictors. These weightings, the Beta weights, indicate the importance of each variable in predicting the criterion.

The technique is valuable when used in vocational and educational psychology, but not so for investigating personality or delinquency. By the method we regress a set of predictor variables into *one* criterion variable, and it does not take much imagination to see that in a complex area of investigation of any sort there is a severe problem in selecting just one single criterion variable. As stated before, criminal behavior is a very complex pattern that involves multiple processes and comprises an enormous variety of offenders and types of offenses. There are myriad forms of delinquency and crime: ultimately as many forms as there are individuals whose impulses and tendencies have been channeled or directed toward illegal behavior. It would be impossible and useless to determine the Beta weights of the variables involved, for they vary as the personalities of the offenders and the different types of illegal behavior vary. All that can be said is what already has been stated: some variables have more weight than others in particular types of illegal behavior, for instance, potential aggressivity for violent crimes or opportunities for some occupational crimes, such as embezzlement.

Association versus Causation

We took great pains in the book, as in the preceding pages, to emphasize repeatedly that we had studied variables and factors that had been found to be *associated* with delinquency and crime. Nevertheless, most reviewers and authors insisted that we implied causation and that our major focus was on mesomorphy (Walters 1973, p. 163; Block 1973, pp. 428–429; Reid 1976, p. 148). It surprises me because the first thing one learns in elementary statistics is that a correlation says nothing about causation. The only paragraph in the book, also expressed in the preceding pages, which could be criticized reads as follows. We were dealing with violent crime and in the section on etiology wrote:

> It is difficult and dangerous to designate the causes of delinquency because research has only discovered *association* or relationships between specific variables and delinquent behavior. Nevertheless, we would advance that the factors or conditions that appear to be closest to the cause of delinquency are: *FAMILY DISRUPTION,* PARTICULARLY WHEN THE CHILDREN ARE MESOMORPHIC. These factors, one environmental, the other constitutional, appear to be very critical and decisive with regard to *violent* and antisocial behavior. (Cortés and Gatti 1972, p. 210)

Perhaps this statement has gone beyond the evidence, but I find it very different from the assertion by Block that "mesomorphy . . . is the cause of delinquent behavior."

In a recent book, after quoting our research, Johnson (1978, p. 150) comments:

> It appears reasonable to believe that a particular set of physical attributes has indirect consequences upon an individual's responses to environmental contingencies. . . . Tall and agile individuals have an advantage in playing basketball. Within our society an individual above average in strength, size and coordination has the advantages in certain occupations, including being a thug.

Such a comment cannot be denied. But then Johnson concludes: "The judgment that certain physical traits go with antisocial behavior adds a sociocultural dimension to the relationship between physiology and temperament in that individuals are pressed to conform to expectations" (Johnson 1978, p. 150). It seems that Johnson has the feeling that any point conceded to biological factors is one point less for social and environmental factors. Even physical attributes add only a sociocultural dimension. Because of the studies quoted in the book (Cortes and Gatti 1972, pp. 71–73) I happen to disagree. The studies on babies, even one week old, and particularly those by Walker (1962, 1963) on 130 boys and girls ranging in age from two to four years, by showing similar relationships between physique and temperament, seem to demonstrate that pressure to conform to expectations is not the only reason, perhaps not even the main reason, for the correlations found between physique and temperament. I believe that nature and nurture always work together, but in the field of temperament, the role of biological and constitutional factors appears to have a slight margin of preponderance.

The Task for the Future

If the assertion that biological factors are inextricably intertwined with psychological and social factors were to be accepted by criminologists and sociologists, the future task in the criminal area should be to progress toward an interdisciplinary approach in order to find the scientific links among such a multiplicity of different but relevant variables.

Rather than measurements on the body or photographs of the body, we urgently need a new approach. I propose that from now on we measure what can be called the *chemotype*. A good start was offered by Williams (1956). Biochemically we are all individuals. We all differ genetically, anatomically, endocrinologically, and neurologically. Even identical twins are different in the composition of their blood or chemotype. We need to investigate all these aspects. In the individual's endowment at birth there seem to be structural, endocrine, metabolic, and neural levels that appear to establish certain characteristics and predispositions. We need to investigate all of them. I believe that different body builds will have different characteristics in the blood's composition. To investigate all this and to deepen our understanding of the causal pathways involved in these aspects and with the multiple psychological and sociocultural variables constantly at work seems to be a challenging task that belongs to all of us and to future generations.

References

Abrahamsen, D. 1960. *The Psychology of Crime.* New York: Columbia University Press.

Adler, F. 1975. *Sisters in Crime.* New York: McGraw-Hill Book Company.

Alsop, S. 1969. "The Coming Holocaust." *Newsweek,* March 3, p. 92.

Argyle, M. 1958. *Religious Behavior.* London: Routledge and Kegan Paul. Also, New York: Free Press, 1959.

Arieti, S. 1967. *The Intrapsychic Self.* New York: Basic Books.

Bandura, A., and Walters, R. 1963. *Social Learning and Personality Development.* New York: Holt.

Barnes, H.E., and Teeters, M.K. 1959. *New Horizons in Criminology.* 3rd ed. Englewood Cliffs, N.J.: Prentice-Hall.

Beeley, A.L. 1954. "A Sociopsychological Theory of Crime and Delinquency." *Journal of Criminal Law, Criminology and Police Science* 45:391–399.

Berg, I. 1967. "Economic Factors in Delinquency." In *Task Force Report: Juvenile Delinquency and Youth Crime.* Washington, D.C.: U.S. Government Printing Office, pp. 305–316.

Biller, H.B. 1970. "Father Absence and the Personality Development of the Male Child. *Developmental Psychology* 2:181–210.

Bloch, H.A., and Geis, G. 1970. *Man, Crime and Society: The Forms of Criminal Behavior.* 2nd ed. New York: Random House. First edition published in 1962.

Block, R. 1973. "Review of Cortés-Gatti's *Delinquency and Crime.*" *Contemporary Sociology: A Journal of Reviews* 2:428–429.

Blum, R. 1969. "Drugs and Violence." In *Crimes of Violence. A Staff Report to the NCCRV.* Vol. 13. Washington, D.C.: U.S. Government Printing Office, pp. 1461–1523.

Blumer, N., and Hauser, P.M. 1933. *Movies, Delinquency and Crime.* New York: Macmillan.

Bolton, R. 1976. "Review of Cortés-Gatti's *Delinquency and Crime.*" *American Anthropologist* 78:148–149.

Briggs, P.F.; Wirt, R.D.; and Johnson, R. 1961. "An Application of Prediction Tables to the Study of Delinquency." *Journal of Consulting Psychology* 25:46–51.

Caldwell, R.G. 1965. *Criminology.* 2nd. ed. New York: Ronald Press.

Cartwright, D.S. 1974. *Introduction to Personality.* Chicago: Rand McNally.

Chateau, J. 1961. "Le milieu professionnel du père et l'equilibre caracteriel des enfants." *Enfance* 1:1-8.

Clark, R. 1970. *Crime in America.* New York: Simon and Schuster.

Cleckley, H. 1976. *The Mask of Sanity.* 5th ed. St. Louis: Mosby.

Cloward, R.A., and Ohlin, L.E. 1960. *Delinquency and Opportunity: A Theory of Delinquent Gangs.* New York: The Free Press of Glencoe.

Cohen, A.K. 1955. *Delinquent Boys. The Culture of the Gang.* New York: Free Press.

———. 1966. *Deviance and Control.* Englewood Cliffs, N.J.: Prentice-Hall.

Cohen, A.K.; Lindesmith, A.R.; and Schessler, K. (eds.) 1956. *The Sutherland Papers.* Bloomington: Indiana University Press.

Cohen, M.R. 1944. *A Preface to Logic.* New York: Holt.

Cole, L., and Hall, I.N. 1966. *Psychology of Adolescence.* 6th ed. New York: Holt.

Conger, J.J.; Miller, W.C.; and Walsmith, C.R.; 1970. "Antecedents of Delinquency: Personality, Social Class, and Intelligence." In P.M. Mussen et al. (eds.), *Readings in Child Development and Personality.* 2nd ed. New York: Harper and Row, pp. 565-588.

Cortés, J.B., and Gatti, F.M. 1972. *Delinquency and Crime: A Biopsychological Approach.* New York and London: Academic Press.

Craft, M. 1960. *Psychopathic Disorders and Their Assessment.* London: Pergamon.

Cressey, D.R. 1969. *Theft of the Nation: The Structure and Operations of Organized Crime in America.* New York: Harper and Row.

Crime in the United States, 1977. See Webster, W.H. (1978).

Davis, J.P. 1966. "The Negro in American Sports." In J.P. Davis (ed.), *The American Negro Reference Book.* Englewood Cliffs, N.J. Prentice-Hall, pp. 775-825.

DeCecco, J.P. and Richards, A.K. 1975. "Civil War in the High Schools." *Psychology Today,* November, pp. 51-56, 120.

Elliot, F.A. 1978. "Neurological Aspects of Antisocial Behavior." In W.H. Reid (ed.), *The Psychopath.* New York: Brunner/Mazel, pp. 149-189.

Elliot, M., and Merrill, E.F. 1950. *Social Disorganization.* New York: Harper.

Empey, L.T. 1978. *American Delinquency.* Homewood, Ill.: Dorsey.

Epps, P., and Parnell, R.W. 1952. "Physique and Temperament of Women Delinquents Compared with Women Undergraduates." *British Journal of Medical Psychology* 25:249.

Erikson, E.H. 1963. *Childhood and Society.* 2nd. ed. New York: Norton.

Eron, L.D. 1963. "Relationship of TV Viewing Habits and Aggressive Behavior in Children." *Journal of Abnormal and Social Psychology* 67:193–196.

Evans, R.I. 1976. *The Making of Psychology.* New York: Knopf.

Eysenck, H.J. 1957. *The Dynamics of Anxiety and Hysteria.* London: Routledge and Kegan Paul.

———. 1966. "Conditioning Introversion-Extraversion and the Strength of the Nervous System." *Proceedings of the XVIIth International Congress of Experimental Psychology,* Moscow, 9th Symposium, pp. 33–34.

———. 1970. *Crime and Personality.* London: Paladin.

———. 1973. *Eysenck on Extraversion.* New York: Wiley.

———. 1978. *The Biological Basis of Personality.* 2nd. ed. Springfield, Ill.: Thomas.

Eysenck, H.J., and Nias, D.R.B. 1978. *Sex, Violence and the Media.* New York: St. Martin's Press.

Ferracuti, F. 1966. *Intelligenza e criminalita: bibliografia.* Milano: Giuffre.

Finney, J.C. 1966. "Relations and Meaning of the New NMPI Scales." *Psychological Reports* 18:450–470.

Fitzpatrick, J.P. 1967. "The Role of Religion in Programs for the Prevention and Correction of Crime and Delinquency." In *Task Force Report: Juvenile Delinquency and Youth Crime.* Washington, D.C.: Government Printing Office, pp. 317–330.

Freud, S. 1910. "The Psycho-analytic View of Psychogenic Disturbance of Vision." In *Standard Edition,* Vol. XI.

Gaddis, T.E., and Long, J.O. 1970. *Killer: A Journal of Murder.* New York: Macmillan.

Gage, N. (ed). 1972. *Mafia, U.S.A.* Chicago: Playboy Press.

Gebhard, P.H.; Gagnon, J.H.; Pomeroy, W.B.; and Christenson, C.V. 1965. *Sex Offenders.* New York: Harper and Row.

Geis, G. 1965. "Statistics Concerning Race and Crime." *Crime and Delinquency* 11:142–150.

———. (ed.) 1968. *White-Collar Crime.* New York: Atherton Press.

Geis, G., and Meier, R.F. (eds). 1977. *White-Collar Crime. Offenses in Business, Politics, and the Professions.* Rev. ed. New York: The Free Press.

Gibbens, T.C.N. 1963. *Psychiatric Studies of Borstal Lads.* London: Oxford University Press.

Gibbons, D.C. 1970. *Delinquent Behavior.* Englewood Cliffs, N.J.: Prentice-Hall.

———. 1977. *Society, Crime, and Criminal Careers.* 3rd. ed. Englewood Cliffs, N.J.: Prentice-Hall.

Glueck, B., 1918. "A Study of 608 Admissions to Sing Sing Prison." *Mental Hygiene* 2:85–151.

Glueck, S., and Glueck, E. 1943. *Criminal Careers in Retrospect.* New York: The Commonwealth Fund.

———. 1950. *Unraveling Juvenile Delinquency.* Cambridge, Mass.: Harvard University Press.

———. 1956. *Physique and Delinquency.* New York: Harper.

———. 1962. *Family Environment and Delinquency.* Boston: Houghton Mifflin.

———. 1968. *Delinquents and Non-delinquents in Perspective.* Cambridge, Mass.: Harvard University Press.

———. 1972. *Identification of Predelinquents.* New York: Intercontinental Medical Book.

Gold, M. 1963. *Status Forces in Delinquent Boys.* Ann Arbor: Institute for Social Research, University of Michigan.

Grossbard, H. 1962. "Ego Deficiency in Delinquents." *Social Casework* 43:71–178.

Hall, C.S., and Lindzey, G., 1978. *Theories of Personality.* 3rd. ed. New York: Wiley.

Hare, R.D. 1968. "Detection Threshold for Electric Shock in Psychopaths." *Journal of Abnormal Psychology* 73:268–272.

———. 1970. *Psychopathy: Theory and Research.* New York: Wiley.

Hare, R.D., and Cox, D.N. 1978. "Psychophysiological Research on Psychopathy." In W.H. Reid (ed.), *The Psychopath.* New York: Brunner/Mazel, pp. 209–222.

Hare, R.D., and Schalling, D. (eds.) 1978. *Psychopathic Behavior: Approaches to Research.* London: Wiley.

Hartmann, H. 1964. *Essays on Ego Psychology.* New York: International Universities Press.

Haskell, M.R., and Yablonsky, L. 1978. *Crime and Delinquency.* 3rd. ed. Chicago: Rand McNally.

Healy, W., and Bronner, E.P. 1936. *New Light on Delinquency and Its Treatment.* New Haven: Yale University Press.

Herskovitz, H.H.; Levine, M.; and Spivack, G. 1959. "Antisocial Behavior of Adolescents from Higher Socio-economic Groups." *Journal of Nervous and Mental Disease* 129:467–476.

Hinderlang, Michael; Hirschi, Travis; and Weis, Joseph. 1979. "Correlates of Delinquency: The Illusion of Discrepancy Between Self-Report and Official Measures." *American Sociological Review* 44(December): 995–1014.

Hirschi, T. 1969. *Causes of Delinquency.* Berkeley and Los Angeles: University of California Press.

Horrocks, J.E., and Gottfried, N.W. 1966. "Psychological Needs and Verbally Expressed Aggression of Adolescent Delinquent Boys." *Journal of Psychology* 62:179–194.

Jenkins, R.L. 1966. "Psychiatric Syndromes in Children and Their Relation to Family Background." *American Journal of Orthopsychiatry* 56:294–300.

Johnson, E.H. 1978. *Crime, Correction, and Society.* 4th. ed. Homewood, Ill.: Dorsey.

Kaplan, S.J. 1958. "Cultural and Community Factors." In J.S. Roucek (ed.), *Juvenile Delinquency.* New York: Philosophical Library.

Karpman, B. 1956. *The Sexual Offender and His Offenses.* New York: Julian Press.

Kaufman, I.; Makkay, E.S.; and Zilbach, J. 1959. "The Impact of Adolescence on Girls with Delinquent Character Formation." *American Journal of Orthopsychiatry* 29:130–143.

Kefauver, E. 1951. *Crime in America.* New York: Doubleday.

Kennedy, R.F. 1960. *The Enemy Within.* New York: Harper.

Kerner Report. See National Advisory Commission on Civil Disorders.

Kiloh, L., and Osselton, J.W. 1966. *Clinical Electroencephalography.* Washington: Butterworth.

Knight, J.A. 1970. *Conscience and Guilt.* New York: Appleton-Century-Crofts.

Konopka, G. 1966. *The Adolescent Girl in Conflict.* Englewood Cliffs, N.J.: Prentice-Hall.

Lefkowitz, M.M, and Cannon, J. 1966. "Physique and Obstreperous Behavior." *Journal of Clinical Psychology* 22:172–174.

Lewis, N.D., and Yarnell, H. 1951. "Pathological Firesetting." *Nervous and Mental Disease Monographs* No. 82.

Lindesmith, A.R. 1968. *Addiction and Opiates.* Chicago Aldine.

Lykken, D.T. 1968. "Neuropsychology and Psychophysiology in Personality Research." In E.F. Borgalta and W.W. Lambert (eds.), *Handbook of Personality Theory and Research.* New York: Rand McNally, pp. 413–509.

Martin, J.M., and Fitzpatrick, J.P. 1965. *Delinquent Behavior.* New York: Random House.

Maslow, A. 1971. *The Farther Reaches of Human Nature.* New York: The Viking Press.

Maurer, D.W., and Vogel, V.H. 1967. *Narcotics and Narcotic Addiction.* 3rd ed. Springfield, Ill.: Thomas.

McCord, W., and McCord, J. 1964. *The Psychopath: An Essay on the Criminal Mind.* Princeton, N.J.: Van Nostrand.

Merton, R.K. 1957. "Social Structure and Anomie (1938)." Reprinted in R.K. Merton, *Social Theory and Social Structure* Rev. ed. New York: The Free Press, pp. 131–160.

Messinger, E., and Apfelberg, B. 1960. "Rapporti esistenti tra comportamento criminale e psicosi, debolezza mentale e tipi di personalita." *Quaderni de criminologia clinca* 3:269–315.

Michaels, J.J. 1961. *Disorders of Character.* Rev. ed. Springfield, Ill.: Thomas.

Miller, W.B. 1970. *City Gangs.* New York: Wiley.

Mischel, W. 1976. *Introduction to Personality.* 2nd ed. New York: Holt.

Monahan, T.P. 1957. "Family Status and the Delinquent Child: A Reappraisal and Some New Findings." *Social Forces* 25:250-259.

———. 1960. "Broken Homes by Age of Delinquent Children." *Journal of Social Psychology* 51:387-397.

Morris, A. 1965. "The Comprehensive Classification of Adult Offenders." *Journal of Criminal Law, Criminology, and Police Science* 56:197-202.

Moynihan, D.P. 1965. *The Negro Family: The Case for National Action.* Washington, D.C.: U.S. Department of Labor.

National Advisory Commission on Civil Disorders. 1968. *Report.* New York: Dutton.

Newman, D.J. 1958. "White-Collar Crime." *Law and Contemporary Problems* 23:735-753.

Nye, F.I. 1958. *Family Relationships and Delinquent Behavior.* New York: Wiley.

Parnell, R.W. 1958. *Behavior and Physique: An Introduction to Practical and Applied Somatometry.* London: E. Arnold.

Parsons, T. 1954. *Essays in Sociological Theory.* Rev. ed. New York: The Free Press of Glencoe.

Pati, P.K. 1961. "Psychopathological Patterns in Delinquency." *Journal of Social Therapy* 7:98-103.

Petri, A. 1967. *Individuality in Pain and Suffering.* Chicago: University of Chicago Press.

Piers, E.V., and Kirchner, E.P. 1969. "Eyelid Conditioning and Personality: Positive Results from Non-partisans." *Journal of Abnormal Psychology* 74:336-339.

Pierson, G.R., and Kelly, R.F. 1963. "HSPQ Norms on a State-wide Delinquent Population." *Journal of Psychology* 56:185-192.

Pittman, D.J. 1958. "Mass Media and Juvenile Delinquency." In J.S. Roucek (ed.) *Juvenile Delinquency.* New York: Philosophical Library pp. 230-247.

———. 1967. *Alcoholism.* New York: Harper and Row.

Pollak, O. 1959. *The Criminality of Women.* Philadephia: University of Pennsylvania Press.

Quay, H.C. 1973. "A Search for Causes." *Contemporary Psychology* 18:274-275.

Quay, H.C. (ed.) 1965. *Juvenile Delinquency: Research and Theory.* Princeton, N.J.: Van Nostrand.

Radzinowicz, L., and Turner, J.W. (eds.). 1944. *Mental Abnormality and Crime.* London: Macmillan.

Radzinowicz, L. (ed.). 1957. *Sexual Offenses.* London: Macmillan.

Rankin, R.J., and Wikoff, R.L. 1964. "The IES Arrow Dot Performance of Delinquents and Nondelinquents." *Perceptual and Motor Skills* 18: 207–210.

Reckless, W.C. 1967. *The Crime Problem.* 4th ed. New York: Appleton-Century-Crofts. 3rd ed. published in 1961.

Redl, F., and Wineman, D. 1952. *Controls from Within.* New York: The Free Press.

Rees, L. 1968. "Constitutional Psychology." In D.L. Sills (ed.), *International Encyclopedia of the Social Sciences,* Vol. 13. New York: Macmillan, pp. 66–76.

————. 1973. "Constitutional Factors and Abnormal Behavior." In H.J. Eysenck (ed.), *Handbook of Abnormal Psychology.* 2nd. ed. San Diego, Calif.: Knapp pp. 487–539.

Reid, S.T. 1976. *Crime and Criminology.* Hinsdale, Ill.: Dryden.

Reid, W.H. (ed.). 1978. *The Psychopath.* New York: Brunner/Mazel.

Reiss, A.J. 1951. "Delinquency as the Failure of Personal and Social Controls." *American Sociological Review* 16:196–208.

————. 1952. "Social Correlates of Psychological Types of Delinquency." *American Sociological Review* 17:710–718.

Report by the National Commission on Violence. See U.S. National Commission.

Report by the President's Commission. See U.S. President's Commission.

Report of the Task Force on the Media. 1969. *Mass Media and Violence.* Washington, D.C.: Government Printing Office. Vol. 9, NCCPV Staff Study Series.

Robin, G.D. 1964. "Gang Member Delinquency: Its Extent, Sequence and Typology." *Journal of Criminal Law, Criminology and Police Science* 55:56–69.

Rodman, H., and Grams, P. 1967. "Juvenile Delinquency and the Family: A Review and Discussion." In *Task Force Report: Juvenile Delinquency and Youth Crime.* Report by the President's Commission. Washington, D.C.: Government Printing Office, pp. 188–221.

Rosenquist, C.M, and Megargee, E.I. 1969. *Delinquency in Three Cultures.* Austin: University of Texas Press.

Ryckman, R.M. 1978. *Theories of Personality.* New York: Van Nostrand Company.

Schachter, S., and Latané, N. 1964. "Crime, Cognition and the Automatic Nervous System." In D. Levine (ed.), *Nebraska Symposium on Motivation.* Lincoln: University of Nebraska Press, pp. 221–273.

Schmauk, F. 1968. "A Study of the Relationship between Kinds of Punishment, Autonomic Arousal, Subjective Anxiety, and Avoidance Learning in the Primary Psychopath." Unpublished doctoral dissertation. Temple University, Philadelphia.

Sellin, T. 1938. "A Sociological Approach to the Study of Crime Causation." In *Cultural Conflict and Crime,* Bulletin 41, chapter 2. New York: Social Science Research Council.

Shaw, C., and McKay, H. 1969. *Juvenile Delinquency and Urban Areas.* Chicago: University of Chicago Press.

Sheldon, W.H. (with the collaboration of S.S. Stevens and W.B. Tucker) 1940. *The Varieties of Human Physique.* New York: Harper.

Sheldon, W.H. (with the collaboration of E.M. Hartl and E. McDermott) 1949. *Varieties of Delinquent Youth.* New York: Harper.

Sheldon, W.H. (with the collaboration of C.W. Dupertuis and E. McDermott) 1954. *Atlas of Men: A Guide for Somatyping the Adult Male at All Ages.* New York: Harper.

Short, J.F. (ed). 1968. *Gang Delinquency and Delinquent Subcultures.* New York: Harper and Row.

Shupe, L.M. 1954. "Alcohol and Crime." *Journal of Criminal Law, Criminology and Police Science* 44:661–665.

Simon, R.J. 1975. *Women and Crime.* Lexington, Mass.: D.C. Heath and Company.

Spergel, I. 1967. *Street Gang Work: Theory and Practice.* Reading, Mass.: Addison-Wesley.

Sutherland, E.H. 1951. "Critique of Sheldon's *Varieties of Delinquent Youth."* *American Sociological Review* 18:142–148. See also in A. Cohen et al. (eds.), *The Sutherland Papers.* Bloomington: Indiana University Press, 1956, pp. 279–290.

———. 1961. *White Collar Crime.* New ed. New York: Holt.

Sutherland, E.H., and Cressey, D.R. 1974. *Criminology.* 9th ed. New York: Lippincott.

Tappan, P.W. 1960. *Crime, Justice and Correction.* New York: McGraw-Hill.

Task Force Report: Juvenile Delinquency and Youth Crime. 1967. Report by the staff of the U.S. President's Commission on Law Enforcement and Administration of Justice. Washington, D.C.: U.S. Government Printing Office.

Thrasher, F.M. 1963. *The Gang.* Rev. ed. Chicago: University of Chicago Press.

Toby, J. 1957. "The Differential Impact of Family Disorganization." *American Sociological Review* 22:505–512.

———. 1967. "Affluence and Adolescent Crime." In *Task Force Report: Juvenile Delinquency and Youth Crime.* Washington, D.C.: U.S. Government Printing Office, pp. 132–144.

Trasler, G. 1962. *The Explanation of Criminality*. London: Routledge and Kegan Paul.

U.S. National Commission on the Causes and Prevention of Violence. 1970. *To Establish Justice, to Insure Domestic Tranquility, Final Report*. Bantam Books. (Official edition, Washington, D.C.: U.S. Government Printing Office, 1969).

U.S. President's Commission on Law Enforcement and Administration of Justice. 1967. *The Challenge of Crime in a Free Society. A Report*. Washington, D.C.: U.S. Government Printing Office.

Walker, R.N. 1962. "Body Build and Behavior in Young Children: I. Body Build and Nursery School Teachers' Ratings." *Monograph of the Society for Research in Child Development*. 27, No. 3, Serial No. 84.

———. 1963. "Body Build and Behavior in Young Children: II. Body Build and Parents' Ratings." *Child Development* 34:1–23.

Wallinga, J.V. 1956. "The Psychopath: A Confused Concept." *Federal Probation* 20:51–54.

Walters, D.R. 1973. "Review of Cortés-Gatti's *Delinquency and Crime.*" *Journal of Criminal Justice* 1:161–163.

Washburn, S.L. 1951. "Review of W.H. Sheldon's *Varieties of Delinquent Youth.*" *American Anthropologist* 53:561–563.

Washburn, W.C. 1963. "The Effects of Physique and Intrafamily Tension of Self-concepts in Adolescent Males." *Journal of Consulting Psychology* 26:460–466.

Webb, M.L. 1943. "Delinquency in the Making: Patterns in the Development of Girl Sex Delinquency in the City of Seattle." *Journal of Social Hygiene* 29:502–510.

Webster, W.H. (ed.) 1978. *Crime in the United States 1977*. Washington, D.C.: U.S. Government Printing Office, October 18.

Weinberg, S.K. 1958. "Sociological Processes and Factors in Juvenile Delinquency." In J.S. Roucek (ed.), *Juvenile Delinquency*. New York: Philosophical Library, pp. 113–132.

West, D.J. 1969. *Present Conduct and Future Delinquency*. New York: International Universities Press.

White, R.W. 1963. *Ego and Reality in Psychoanalytic Theory*. Psychological Issues II. New York: International Universities Press.

Wilkins, L.T. 1960. *Delinquent Generations*. London: Her Majesty's Stationery Office, Home Office Research Unit.

Williams, J. 1956. *Biochemical Individuality*. New York: Wiley.

Winslow, E.A. 1939. "Relationships Between Employment and Crime Fluctuations as Shown by the Massachusetts Statistics." In *National Commission on Law*, Washington, D.C.: U.S. Government Printing Office, vol. 1, pp. 257–333.

Wirth, L. 1931. "Culture Conflict and Delinquency." *American Journal of Social Forces* 9:484–492.

Wolfenstein, M., and Leites, N. 1950. *Movies: A Psychological Study*. New York: The Free Press of Glencoe.

Wolfgang, M.E. 1958. *Patterns in Criminal Homicide*. Philadelphia: University of Pennsylvania Press.

Wolfgang, M.E., and Ferracuti, F. 1967. *The Subculture of Violence*. London: Tavistock.

Yablonsky, L. 1970. *The Violent Gang*. Rev. ed. Baltimore: Penguin Books. (Published originally in 1962.)

Yochelson, S., and Samenow, S.E. 1976. *The Criminal Personality*. Vol. I: *A Profile for Change*. New York: Aronson.

Zamorano, M. 1961. *Hacia el conocimiento del crimen*. Santiago, Chile: Imprenta Cultura.

11

Genetic and Social Factors in Human Behavioral Disorders

Kenneth K. Kidd

The purpose of this chapter is not to present any general conclusions—none are possible—but rather to give some specific examples that illustrate the problems involved in trying to understand the confounding of genetic and nongenetic factors in the etiology of human behavioral disorders.

One background fact is that virtually any disease that one might think of, including behavioral disorders, shows a familial concentration. A second fact is that most disorders, certainly all the behavioral disorders to be considered here, show clear evidence of environmental modification. Therefore, from the very beginning it is quite clear for many disorders that a familial concentration exists and that nongenetic factors are very important in determining how the disorder is expressed and, in some cases, even in determining whether it is expressed. As a geneticist I would like to jump to the conclusion that a familial concentration is evidence for the existence of some underlying genetic factors; however, genetic and environmental similarity are so highly correlated that the mere fact that individuals are related biologically implies, on average, that they have much more similar environments than do individuals who are unrelated. In addition, as discussed by Feldman and Cavalli-Sforza (chapter 3 in this volume) some aspects of culture and environment can be transmitted through several generations. Consequently, the fundamental question that has to be asked is not whether a disorder is caused by nature or by nurture, but *to what degree and in what manner is the variation among individuals determined by genetic variation among individuals.* This phrasing of the question stems from a primary interest in genetics, but the other half of that question is to what degree and in what manner is the variation determined by variation in environmental or cultural factors. Neither can be adequately answered without considering both genetic and nongenetic factors.

One goal of studies of behavioral disorders is an understanding of the etiology and the development of these disorders. Therefore, some ancillary questions also have to be considered, including the question of heterogeneity. From a human genetics perspective it is known that most diseases of largely genetic origin have been shown to be heterogeneous with sometimes

This work was supported in part by grants NS 11786 and MH 28274.

many different individual, but distinct, genetic causes. For any disorder with genetic heterogeneity the genotype–environment interactions may be different in the different genetic types. The problem of disentangling several different interactions is too complex at present because the question of genetic and social factors in human behavioral disorders is too poorly formulated.

Specific Cases of Gene–Environment Interaction

The problem must be approached initially in a relatively simple and very specific manner. Three examples will be used to illustrate approaches to gene-environment interaction: ankylosing spondylitis, stuttering, and Gilles de la Tourette syndrome. Ankylosing spondylitis is a purely medical disorder with little in terms of behavior as part of its onset. However, it illustrates some genetic conclusions that are very relevant to the study of behavioral disorders, and demonstrates very significant consequences of the sex of the individual on the disorder. The second disorder is stuttering, a very specific speech disorder being studied as a model for many other behavioral disorders. Last, pilot data on Gilles de la Tourette syndrome will be discussed. These data show that this syndrome, which has classically been considered a neurosis and only very recently been considered to be neurological, has very similar familial patterns to those seen for stuttering. Some of the conclusions and the logic used in analyzing stuttering can be transposed virtually intact to Gilles de la Tourette syndrome. For none of these disorders do we have final answers but progress toward a clearer understanding is significant.

Ankylosing Spondylitis

This disorder is a progressive rheumatoid arthritis of the spine with an onset in the late teens or early adulthood. The disorder may not progress very far but it can progress to fusion of most vertebrae of the spine, essentially immobilizing the spine. It has classically been considered a disorder occurring almost exclusively in males; in most clinics the sex ratio is about 9 males to every female. Within the last ten years researchers have found a very strong population association of this disease with a particular antigen, B27, at the HLA system (Brewerton et al. 1973; Schlosstein et al. 1973). (The human lymphocyte antigen system, the major histocompatibility system in man, mediates in a very complex way antibody responses to foreign antigens and is primarily responsible for rejection of tissue grafts between nonidentical individuals.) While the population association is both highly significant and large in magnitude, it cannot by itself really explain how the genes of the HLA system might be involved in the disease process. However, through family studies using HLA linkage a great deal has been learned.

In a family study of a disorder one generally starts with an affected individual, the *proband,* and then studies the relatives. In studies of ankylosing spondylitis (Kidd et al. 1977) all probands had the B27 antigen of the HLA system (as most patients do have) and that chromosome was followed among the relatives, whether it was transmitted from a parent to the proband, transmitted by the proband to his offspring, or independently transmitted to siblings. The relatives were then classified quite unambiguously into those that have this chromosome, or at least the segment marked by the genes for the HLA antigens, and those that do not have this segment of the chromosome. Among the relatives who did not have this marked chromosome there were no cases of ankylosing spondylitis found. However, among the relatives that did have this chromosome, 38 percent were affected with some stage of the disease—very mild in some, more severe in others. Clearly, genetic interaction with some other factors occurs because the B27 antigen is not sufficient for the development of the disease. In fact, it is not clear that the antigen B27 itself is important; the antigen simply serves as a marker for several hundred genes that are inherited in the region of the chromosome around the HLA gene that determines the antigen. However, whatever nongenetic factors may be involved in precipitating the illness apparently have no effect in the absense of the necessary gene. We do not yet know what those nongenetic factors are, but one hypothesis is that certain bacterial infections trigger off an autoimmune reaction in susceptible individuals.

Because of our increased genetic understanding, we now have a much better research paradigm for identifying the relevant nongenetic factors. Presumably all the relatives carrying the B27 antigen are susceptible from a genetic point of view, that is, they have the necessary predisposing gene. These genetically predisposed relatives can then be divided into those that are affected and those that are not affected, and the medical histories of the two groups can be compared. We can look at antibody titers of various pathogens or we could even do a prospective study. Studying all first-degree relatives would add a great deal of noise to the system; significant findings may be impossible because individuals who are not susceptible would also be included; in nonsusceptible individuals the presence or absence of the relevant environmental factor would actually be irrelevant.

Studying the relatives for illness has also shown that the sex difference appears not to be real; it seems to be an artifact of ascertainment. Among the relatives the females with the B27 antigen had the same risk, approximately 40 percent, of developing ankylosing spondylitis as did the males

Table 11–1
Frequency of Ankylosing Spondylitis among Relatives of Patients According to Sex

		Primary Cases	
		Males (n = 33)	Females (n = 8)
All first-degree relatives	Males	9/41 = 0.22 ± 0.06	3/11 = 0.27 ± 0.13
	Females	11/42 = 0.26 ± 0.07	1/14 = 0.07 ± 0.07

Source: Kidd et al. 1977.

Note: All first-degree relatives are considered in the denominator. No individual lacking the B27 allele present in the proband was affected. Values given as number affected/total = mean ± standard error (3 d.f.) = 2.3 $p = 0.50$

with the B27 antigen (Table 11–1). The males, however, were more severely affected. Thus, three factors probably account for the classic sex ratio reported in clinic populations: (1) In males the disease probably progresses further and becomes more severe and, hence, males are slightly more likely to be seen in a clinic and to have an unambiguous diagnosis. (2) The social roles of males have, at least in the past, made limited mobility of the spine more economically critical, and hence, made males more likely to seek medical aid. (3) Almost certainly part of the sex difference is iatrogenic—the medical literature says that females are almost never affected with ankylosing spondylitis and, hence, a doctor is much more reluctant to make the diagnosis in a female; a female with lower back pain may be diagnosed as having rheumatoid arthritis, not ankylosing spondylitis, or "just female troubles."

Because of the genetic design and the fact that a genetic perspective has been followed, social role differences and the sex effect have been shown to be important for a medical disorder. The same design provides the potential for a much greater understanding of the nongenetic factors involved in the onset of the disease.

Stuttering

The same firm conclusions and the same reasonably clear understanding are not yet possible for stuttering, but some significant progress has been made for this disorder. The most useful definition is that stuttering is a temporal disruption in the simultaneous and successive programming of muscular movements required to produce one of a word's integrated sounds. It is basically a disruption in the timing of speech, in the smooth flow of speech.

This disruption is characterized by a block that may be manifest as repetition, as a prolongation, or sometimes as a silent gap in the middle of a word. In contrast to some others this definition represents a phenomenological diagnostic approach; it is not based on abstract theory.

Some background on stuttering may help clarify the disorder. The onset is always in childhood, starting from the onset of speech. The fiftieth percentile for onset is at around age 5; the ninetieth percentile is reached by age 7, the ninety-ninth percentile by age 9. There are isolated reports of onset after 12, but an extremely small number of such late onsets. Recovery is a major phenomenon: most stutterers recover before they become adults. Those that persist in stuttering into adulthood apparently cannot recover normal spontaneous fluent speech. Adult stutterers can learn to control their speech, but it is generally through a conscious mechanism of speech monitoring, not through the automatic system that all of us generally use. Stuttering also has a very marked sex effect; the sex ratio among stutterers is about 4 to 5 males to every female.

There have been many different hypotheses on the origin of stuttering. One that had a great deal of influence starting in the early 1940s and continuing through the early 1950s was the semantogenic hypothesis (Johnson 1959). This hypothesis is based on the observation that all children normally have some difficulty with speech as they are learning to talk. According to Johnson, the negative reactions of parents and others in the child's environment to these disfluencies create the speech problem. When the parent or community diagnose the speech disfluency as stuttering, the child becomes a stutterer because he was labeled a stutterer. This hypothesis proposed a purely social etiology for stuttering, but it is now not considered valid for a variety of reasons. Most recent researchers into stuttering consider stuttering to be basically an organic disorder (Van Riper 1971; Bloodstein 1975). In so doing they are separating quite clearly the presence or absence of the basic problem in timing of speech from the severity of the stuttering. This is necessary because of the contextual aspects of stuttering. If a stutterer is anxious or nervous, his stuttering is much worse. The social context may also affect the expression of symptoms: in a classroom his stuttering may be very severe, but out with friends drinking in the evening his stuttering may be almost nonexistent. These contextual factors affecting short-term temporal variation are also likely to be distinct from factors affecting the persistence of stuttering into adulthood as well as from factors determining the concomitant behaviors.

While little is yet known about recovery versus persistence into adulthood, the concomitant behaviors are better understood. These behaviors may include eye blinks, grimaces, and shoulder jerks. These concomitant behaviors are believed to develop through a behavioral learning process. If a stutterer gets into a block, a distraction will often get him past it, allowing him to proceed with his speech. That distraction can be external or it may be generated by the stutterer, like blinking his eyes, etc. These distractions will occasionally work, as long as they are unusual; they become learned habits with intermittent reinforcement. Once a specific action becomes a habit it becomes ineffectual, but there is no mechanism for unlearning it. Severe stutterers can develop a whole repertoire of concomitant behaviors that are not part of the basic disorder but that are the result of behavior and reinforcement.

In some work the familial nature of stuttering is explained as the result of the attitudes and traits of parents also running in families (Gray 1940; Johnson 1959). Thus, traditionally there has been the realization that stuttering runs in families, but this was attributed to cultural attitudes and parental traits also running in families. Such cultural transmission does not depend on the semantogenic hypothesis; other forms of parent-child interaction could very well be involved. Some of the characteristics of parents of stutterers that have been suggested by numerous investigators include the setting of high goals and standards of performance, high expectations and evaluations of speech, covert forms of parental rejection, maternal overprotection and oversupervision, withholding of approval for accomplishments, an unwillingness to encourage independence, submissiveness and low social dominance on the part of the parents. I will come back to these later.

There are several reasons, a priori, for thinking that stuttering might have a genetic component. First, completely apart from its familiarity, which is well documented, stuttering is a very specific disorder. Stuttering itself is what runs in families, not speech disfluency in general. There are many other types of speech disfluence. Cluttering is one that can show a pattern of autosomal dominant inheritance (Op't Hof and Uys 1974) and is characterized by speech so rapid that whole syllables and words get lost and left out—the stream of sound becomes unintelligible. Cluttering can be improved by telling the clutterer to slow down, to think about what he is saying. If the clutterer is speaking to a superior or is a little more anxious and concerned about his speech, the speech improves. In the same situations the speech of a stutterer becomes worse. Thus, in terms of symptoms and reactions to situations, stuttering and cluttering are quite different. Articulation disorders are other speech defects that run in families and that are confounded with stuttering.

A second reason for believing that stuttering might be genetic, aside from its being a specific disorder, is that one would expect there to be genetic defects in speech. Speech is the newest, most uniquely human form of behavior. Higher primates have language. Chimpanzees have been taught to communicate with abstract symbols using a simple form of grammar (Bourne 1977). Admittedly, the communication is not as advanced as human grammar but it is, nonetheless, a grammar. However, chimpanzees do not have speech; only humans have speech. Speech is an extremely complex form of behavior. Over one hundred muscles are involved in speech, and the movements from the production of one sound to the production of the next are extremely complex and require a sophisticated servomechanism involving complex feedback and monitoring. It is estimated that approximately 140,000 neuromuscular events per second are involved in the normal production of speech (Bateman 1977). That does not count the thought processes that are going on ahead of the speech, assuming that they are present. Thus, though humans are genetically difficult to study, there is no animal model of stuttering.

Stuttering does not show a simple pattern of inheritance in families. Sample pedigrees are given in figure 11-1. Sometimes a parent stutters.

FAMILY 22

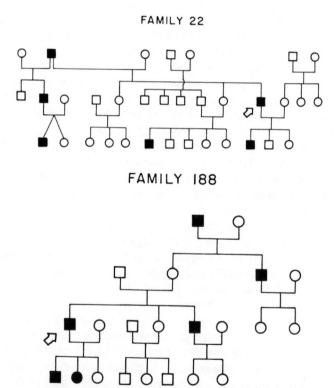

FAMILY 188

Squares represent males; circles represent females. Darkened symbols indicate an individual who ever stuttered; arrows indicate the probands. Most pedigrees had fewer stutterers than these, but these show the irregular patterns more clearly.

Figure 11-1. Two Selected Pedigrees from a Total Dataset of Over 600 Families

Sometimes nobody else in a reasonably large pedigree has ever stuttered. Some pedigrees look almost as though stuttering were due to an X-linked recessive because a male stutterer will have an unaffected sister who has affected sons. However, in many pedigrees we find apparent male-to-male transmission, which argues against X-linked inheritance. Although there is always a chance that the mother was a carrier, pedigrees with an affected father and an affected son are found in sufficient numbers that we can confidently exclude an X-linked component to the inheritance. However, inheritance of stuttering is clearly sex modified. Males are more frequently affected than females, both among probands and among relatives.

Both the sex effect and the high familial concentration can be seen in table 11-2, which summarizes data on 511 families. Consider first the relatives of male probands. Among the male relatives approximately 20 percent report having ever stuttered at some time in their lives, whether they had recovered before they became adults or whether they still stutter in adulthood. Among the female relatives, however, the frequency of having ever stuttered is much lower, on the order of 5-10 percent. Interestingly, among relatives of female probands, the same qualitative pattern exists, but the frequencies are systematically higher. Thus, there appears to be a direct effect of sex on how frequently a relative stutters. Statistically the effect is highly significant. Contrasting the frequencies in table 11-2 with the findings for ankylosing spondylitis, shown in table 11-1, emphasizes that for stuttering the sex effect is a very real phenomenon.

One of the major questions to be asked of these data is whether or not transmission exists: Does a parent stuttering increase the risk of an offspring stuttering? The analysis used to test this question starts by dividing the families into subsets. Families of adult probands were first classified by whether the proband was a male or a female. They were also classified by stuttering in the proband's parents into four types of families: (1) neither parent ever stuttered, (2) father stuttered, (3) mother stuttered, (4) both parents stuttered. Table 11-3 gives the frequencies of stutterers among the remaining relatives, especially the brothers and sisters of our probands, according to these classifications. There were very few families in which the mother stuttered or both parents stuttered and, hence, they have been omitted from table 11-3. There is a marked and statistically significant increase in stuttering if the father of the proband also stuttered. This same pattern also occurs among relatives of female probands (table 11-4). Comparing siblings of probands who had two parents who did not stutter with those whose father stuttered, the frequency among brothers of female probands goes from 18 percent up to 33 percent and among sisters, from 9 percent to 17 percent. Several elaborate analyses show that this nonrandom distribution within families is statistically significant in a pattern that is a clear demonstration that stuttering is transmitted within families (Kidd and

Table 11-2
Stuttering among Relatives of Stutterers
(number ever stuttered total relatives)

	Male Probands[a]	Female Probands[b]
Fathers	$81/384 = 0.211 \pm 0.021$	$28/127 = 0.220 \pm 0.037$
Brothers	$99/480 = 0.206 \pm 0.018$	$28/130 = 0.215 \pm 0.036$
Sons	$28/122 = 0.230 \pm 0.038$	$20/51 = 0.390 \pm 0.068$
Mothers	$23/384 = 0.060 \pm 0.012$	$16/127 = 0.126 \pm 0.029$
Sisters	$20/412 = 0.049 \pm 0.011$	$15/153 = 0.098 \pm 0.024$
Daughters	$11/110 = 0.100 \pm 0.029$	$7/44 = 0.159 \pm 0.055$

[a]$n = 384$; [b]$n = 127$.

Table 11-3
Risk to Relatives of Adult Male Stutterers

	Family Type	
	I ($N = 227$, 78%)	II ($N = 49$, 17 %)
Brothers	$.174 \pm .022$	$.236 \pm .057$
Sisters	$.022 \pm .009$	$.100 \pm .039$
Sons	$.214 \pm .041$	$.350 \pm .107$
Daughters	$.093 \pm .029$	$.095 \pm .064$

Source: Kidd and Records (1979).

Table 11-4
Risk to Relatives of Adult Female Stutterers

	Family Type	
	I ($N = 69$, 70%)	II ($N = 19$, 19%)
Brothers	$.176 \pm .044$	$.333 \pm .096$
Sisters	$.088 \pm .030$	$.167 \pm .076$
Sons	$.333 \pm .075$	$.714 \pm .171$
Daughters	$.125 \pm .058$	$0 \pm .152$

Source: Kidd and Records (1979).

Records 1979; Kidd et al. 1981). Some of these analyses do not assume a genetic hypothesis but simply demonstrate that if a parent also stuttered the risk to other siblings increases. That is not by any means proof that the transmission must be genetic, but it is clear that there is transmission through these families of something that is related to stuttering. Whatever is transmitted also interacts with the sex of the individual: it takes more of it to make a female stutterer. Conversely, if one starts with a female stutterer, there are more of these transmitted factors in her family and hence a higher risk to other relatives (Kidd et al. 1978).

The next questions consider how to explain the transmission. Genetic models might be used to explain the transmission. Stuttering does not follow a clear single-gene pattern but it may very well be explained by some threshold model in which a major locus segregates three genotypes but nongenetic variation modifies the risk of stuttering for each genotype. A threshold then determines who does and who does not stutter. We could have heterozygotes who stutter if they have a sufficiently stressful environment, and we could have homozygotes who do not stutter if they have a significantly benign or ameliorating environment. The position of the threshold and the gene frequencies can be varied to find the best explanation of the data. For stuttering the results are given in table 11–5. This solution explains the family data quite well with all of the transmission being genetic. The gene frequency in the solution is 4 percent, not a rare gene, but not a very common one either. Most individuals in the total population lack the gene. About 8 percent of the population is heterozygous and less than 2 in 1,000 are actually homozygous. The penetrance—the probability that an individual with a particular genotype will ever stutter, as a function of sex— is virtually zero for the normal genotype; it is one (or 100 percent), irrespective of sex, for homozygotes for the stuttering allele. A striking sex difference is found for heterozygotes: penetrance is about 40 percent for a heterozygous male but only 11 percent for a heterozygous female. These parameter values predict a lifetime prevalence of around 4 percent of the population of males and around 1 percent of the population of females, about the values we know for the lifetime prevalence of ever stuttering at some time in childhood. Finding an acceptable solution does not constitute proof that a single locus is responsible for stuttering, but it is highly suggestive. Especially noteworthy is that the model does not require the interaction of genetic and nongenetic factors be confined to heterozygotes, and yet biological principles would suggest that genotype as the one in which that interaction is most likely to be significant.

Stuttering is transmitted, and the transmission can be explained by a genetic model. Can it also be explained in a quantitative way by a cultural transmission model? The simplest of the cultural transmission models is one in which the behavior is mimicked—a child stutters because he mimics stut-

Table 11–5
**Best Fit of the Single Major Locus Model to the Family Incidence Data
on Stuttering**

Stuttering allele (S) frequency = 0.040 ± 0.007
Predicted general incidences: 0.035 for males; 0.010 for females
Penetrances for each genotype and sex

	Genotypes		
	NN	NS	SS
Male	0.005 ± 0.003	0.378 ± 0.025	1.0
Female	0.0002 ± 0.0002	0.107 ± 0.019	1.0

The goodness of fit $X_3^2 = 4.65$, $p = 0.22$

Source: Kidd and Records (1979).

tering behavior in a parent. That hypothesis can be quite convincingly rejected because it cannot account for much of the transmission (Kidd et al. 1978). Of those probands who had a parent who stuttered, it was the father who stuttered in most cases; over half of the fathers who ever stuttered had recovered before they became adults. Thus, in about 90 percent of all families, at the time the proband was born neither parent stuttered so that no stuttering model existed. In total, only a very small percentage of all stutterers had any stuttering model they could mimic.

As mentioned earlier, many other hypotheses about the behavioral traits of the parents have been suggested. Might some of those traits be culturally transmitted? In light of what is now known about the family pattern of stuttering, it is interesting to go back and reevaluate some of those reports. Remember that on average about 20 percent of the stutterers had a father who had at some time stuttered. In one study (Johnson 1959) a question was asked about how often the parents corrected their child's pronunciation. A significantly greater proportion of mothers of stutterers responded "never" than did mothers of controls—about 25 percent of the mothers of stutterers versus 7 percent of the mothers of controls. This was interpreted as mothers of stutterers being more permissive and having lower social dominance. The fathers displayed a similar trait, 21 percent versus 13 percent never corrected their child for pronunciation. Unfortunately, this study provided no information on whether or not those fathers (or mothers) who never corrected the child's speech had ever stuttered when they were children; to interpret these results that is an additional fact we would like to know. If 21 percent of the fathers had themselves at some time been stutterers, might not they be expected to have different behavioral attitudes toward their child's stuttering? Knowing whether a parent ever stuttered is an important factor in understanding the behavioral attitudes and possibly the etiology. Since we now know that transmission exists and that there is a high familial frequency of stuttering, we must reexamine all the psychosocial studies. None of them has ever taken into account whether or not the parent being studied had ever stuttered. We cannot yet conclude what is involved in stuttering, but we can more clearly state what needs to be done.

Gilles de la Tourette Syndrome

The syndrome of Gilles de la Tourette is another behavioral disorder that has its onset in youth, has quite variable manifestations over time, and is characterized by speech defects. Sometimes the speech symptoms are like stuttering, but more frequently the speech symptoms are multiple vocals tics and strange barking noises: these individuals will make strange sounds irregularly and without any clear relationship to speech context. The most intriguing symptom, present in nearly half the patients, is coprolalia, which is involuntary, obscene speech or swearing uttered at inappropriate times and places. From a psychoanalyst's point of view this seems to be a classic manifestation of repression. More recently it has become clear that the disorder runs in families and can be treated with drugs (Shapiro et al. 1978).

These findings suggest that Gilles de la Tourette syndrome may be a neurological disorder. Additional evidence for its being a neurological disorder is a class of aphasias in which individuals lose the ability to speak voluntarily; these patients cannot, on desire, manage to speak. However, if they get angry, they can swear fluently. Thus, all the speech production mechanism and knowledge of words are still present. Language that is initiated under high emotional stress or with high emotional quality to the words appears to have a different initiation point separate from that part of the brain that initiates normal speech. This neurological finding suggests that the involuntary swearing in Gilles de la Tourette syndrome may be a minor epileptic-like seizure in that part of the brain that triggers off highly emotional speech. In this sense coprolalia may be seen as another symptom similar to the multiple tics that indicate some irregularity in motor coordination.

We have recently collected some family data on Gilles de la Tourette syndrome. Because these are pilot data, the numbers are small and most results are at borderline statistical significance. Analyses have been done in much the way as analyses of the data on stuttering (Kidd et al. 1980). Affected males are more common than affected females. The pattern within families shows 17–20 percent of male relatives affected, 10 percent of female relatives affected. Also, among relatives of female patients the frequencies of affected relatives are higher. Though the specific frequencies are different, the overall familial pattern is virtually identical to the familial pattern of stuttering. Gilles de la Tourette syndrome evidently has a form of transmission that interacts with the sex of the individual such that whatever is transmitted, it takes more of it to make a female affected but given an affected female, there are consequently more of these transmitted factors in the family. Thus, a disorder that was for many years considered a purely psychosocial phenomenon is now thought of as basically a neurological disorder. Our understanding is still incomplete, and psychosocial factors undoubtedly interact in the production from the basic neurological disorder to the specific symptom pattern of the individual patient.

Conclusion

The examples discussed above show that it is necessary, when studying human behavioral disorders, to adopt a research paradigm that incorporates the family in a way that at least allows for genetic analyses or for analyses of the familial patterns like those that have been done for stuttering and Gilles de la Tourette syndrome. For a specific disorder one may find that there is no genetic component, that there is no evidence for transmission, that purely random cultural factors explain the pattern. That, in itself, would be a significant finding. If, however, one finds evidence for transmission, one can have a much clearer understanding and much greater power for identifying what might be the individual environmental, social, or genetic factors involved.

References

Bateman, H.E. 1977. *A Clinical Approach to Speech Anatomy and Physiology.* Springfield, Ill.: Thomas.

Bloodstein, O. 1975. *A Handbook on Stuttering,* Revised edition. Chicago: National Easter Seal Society for Crippled Children and Adults.

Bourne, G.H. 1977. *Progress in Ape Research.* New York: Academic Press.

Brewerton, D.A.; Caffrey, M.; Hart, F.D.; James, D.C.O.; Nicholls, A.; and Sturrock, R.D. 1973. Ankylosing spondylitis and HL-A 27. *Lancet* 1:904–907.

Feldman, M., and Cavalli-Sforza, L.L. 1982. Darwinian selection and behavioral evolution. Chapter 3 in this volume.

Gray, M. 1940. The X family: A clinical and laboratory study of a "stuttering" family. *J. Speech Hear. Disord.* 5:343–348.

Johnson, W. 1959. *The Onset of Stuttering. Research Findings and Implications.* Minneapolis: Minnesota Press.

Kidd, K.K.; Bernoco, D.; Carbonara, A.O.; Daneo, V.; Steiger, U.; and Ceppelini, R. 1977. Genetic analysis of HLA associated diseases: The "illness susceptible" gene frequency and set ratio in ankylosing spondylitis. In Dausset, J. and Svejgaard, A., (eds.), *HLA and Disease.* Copenhagen: Munksgaard, pp. 72–80.

Kidd, K.K.; Heimbuch, R.C.; and Records, M.A. 1981. Vertical transmission of susceptibility to stuttering with sex-modified expression. *Proc. Nat. Acad. Sci. USA* 78:606–610.

Kidd, K.K.; Kidd, J.R.; and Records, M.A. 1978. The possible causes of the sex ratio in stuttering and its implications. *J. Fluency Disorders* 3:13–23.

Kidd, K.K.; Prusoff, B.A.; and Cohen, D.J. 1980. Familial pattern of Gilles de la Tourette syndrome. *Arch. Gen. Psychiatry* 37:1336–1339.

Kidd, K.K., and Records, M.A. 1979. Genetic methodologies for the study of speech. In Breakefield, X.O. (ed.), *Neurogenetics: Genetic Approaches to the Nervous System.* New York: Elsevier-North Holland, pp. 311–344.

Op't Hof, J., and Uys, I.C. 1974. A clinical delineation of tachyphemia (cluttering): A case of dominant inheritance. *S. Afr. J. Med. Sci.* 48: 1624–1628.

Schlosstein, L.; Terasaki, P.I.; Bluestone, R.; and Pearson, C.M. 1973. High association of an HL-A antigen, W27, with ankylosing spondylitis. *New Engl. J. Med.* 288:704.

Shapiro, A.; Shapiro, E.; Brunn, R.; et al. 1978. *Gilles de la Tourette Syndrome.* New York: Raven Press.

Van Riper, C. 1971. *The Nature of Stuttering.* Englewood Cliffs, N.J.: Prentice-Hall.

12 The Impact of Psychopharmacology on the Mental-Health-Service System

Gerald L. Klerman
and *Gail Schechter*

Interactions between biological and psychological influences on behavior are most intense in the area of mental illness. Advances in biological and psychosocial treatments have had a major impact on the mental-health-service system. This chapter attempts to lessen gaps in the knowledge of social scientists about advances in mental-health treatments.

Within the past twenty years there have been dramatic advances not only in drug therapies but also in social attitudes and psychosocial treatments. These have radically transformed the structure of the mental-health-service system, particularly inpatient hospitalization. At the same time, the widespread use of minor tranquilizers has had a major impact on the attitudes and expectations of the public regarding the nature of stress and the mechanisms used to cope best with it. These advances and the resultant controversies are important contributions to a better understanding of the interplay between biological and social determinants of behavior.

Historical Events Affecting the Mental-Health-Service System

The mental-health-service system, until the 1950s, was monopolized by large public mental institutions. Since that time, there has been a dramatic shift in the locus of treatment from institutional to out-patient settings and a restructuring of mental health services to accommodate this shift.

Changes in the mental-health-service system during the past quarter century are in major part the result of new technologies—both biological and psychosocial. The advent of psychopharmacological treatments, with the introduction of chlorpromazine in the early 1950s, is often considered the main event that revolutionized the treatment of mental illness (Swazey 1974), although psychosocial therapies and theories were making significant contributions during the same period.

In the twenty-five years since their introduction, there has been a proliferation of psychotropic drugs used in the treatment of mental disorders. A large number of well-designed, documented studies (Levine et al. 1971) provide compelling evidence demonstrating the efficacy of psychotropic agents for treating schizophrenia, affective disorders, and anxiety. The application of sophisticated research methodology, especially controlled clinical trials using random assignment of subject, placebo controls, and double-blind techniques, contributes greatly to the strength of the evidence. Using these research designs, it has been conclusively demonstrated that psychotropic drugs are significantly more efficacious than placebo, not merely related to expectations, or not the result of the Hawthorne effect (that is, when the process of measurement and the observation influences and changes what is being measured) (Rosenthal and Rosnow 1969).

The scientific methods for demonstrating that an intervention does work, however, are different from the investigation required to demonstrate how it works. In the 1950s and 1960s, although the efficacy of psychotropic drugs was established, their modes of action were not understood. Ideally, scientific advances should progress from basic laboratory discoveries to clinical applications. However, this ideal was not manifested during the development of the first wave of psychotropic drugs; clinical effects were discovered first and they, in turn, stimulated the development of basic research to explain the therapeutic efficacy of these drugs (Klerman 1978).

It is common in the history of medicine for a treatment to work even though knowledge as to mode of action is unavailable. For example, aspirin works to alleviate fever, but the mechanism responsible for that effect is not clear—what part of the brain or how the white cells are involved.

When psychoactive drug treatments for the mentally ill were introduced, serious efforts had been already underway to generate and apply new psychosocial technologies. Beginning in the late 1940s and early 1950s, social attitudes and hospital practices began to change, particularly among administrators of mental institutions, resulting in therapeutic reforms, especially greater willingness to discharge patients. These technologies included group techniques, use of nonrestraint, open-door policies, and other procedures and methods. These approaches had three distinct characteristics:

1. they were psychosocial, not biological or pharmacological;
2. they were directly and intentionally related to social science and historical research and theory; and
3. their openly declared goals included not only the improvement of individual patient care and treatment but also the intent to reform and reconstruct the social organization and dynamics within the mental hospital.

These psychosocial reforms were partially spurred by the many criticisms of public mental institutions. A number of studies identified the custodial, authoritarian, and bureaucratic nature of mental hospitals that served to undermine their publicly mandated therapeutic goal and the conscious intent of their leadership and staff. Stimulated by these studies and by growing public concern, major efforts at therapeutic reform were widely initiated in the mental-hospital field during the 1950s and 1960s, although the British "open hospital" experience began in the late 1940s.

Psychosocial reform and rehabilitation efforts quickly affected the mental-health professions and were crystallized in theory and research as "social psychiatry" and in practice as "therapeutic community." Within mental hospitals in the United States and Great Britain, progressive superintendents and their staffs opened doors and eliminated restraints. They also took steps to decrease the social distance between patients and staff, to facilitate communication among staff groups, and to upgrade employee morale and training. These intra-hospital reforms also led to experimentation with various alternatives to hospitalization, such as halfway houses and crisis intervention centers, the forerunners of many of today's community mental-health and corrections innovations.

These reforms were idealized in the concept of the *therapeutic community* enunciated by Maxwell Jones (1953). Jones united the spirit of egalitarian democracy with the techniques of group dynamics to level unequal status, to facilitate communication, and to share decision-making throughout the total institution. The therapeutic community became the rallying slogan of the mental-hospital reform efforts of the 1950s. Initially developed in Great Britain for the treatment of individuals with personality disorders, the therapeutic community was rapidly extended in the United States to inpatient settings for acute psychotics and later to the treatment of adolescent drug addicts in residential settings. The therapeutic-community movement within mental-health institutions has parallels in the innovative outpatient psychotherapy and counseling proposed in recent years.

As Gruenberg (1966) has pointed out, these new "technologies" had impact on chronic as well as on acute patients. The new psychosocial techniques improved care for people with chronic, severe mental disorders and often changed the organization of treatment into a pattern called *community care.* The anticipated benefit of this new technology was the prevention of chronic deterioration in personal and social functioning.

Unfortunately for historical analysis, the emergence of these new psychosocial treatments for the mentally ill occurred simultaneously with the advent of psychopharmacology. One aspect of the dramatic development of the psychopharmacological approach was that the psychosocial reform efforts did not gain as much public attention as did the new drug treatments. Some community mental-health programs already were operating when the first new drug was discovered in France in 1952. In the mid-1950s the United States began to import the new drugs from France and new psychosocial techniques from Britain. Because the new drugs made therapeutic community care practices easier, some observers concluded that all subsequent changes in treatment and outcome had been caused by the drugs.

Both the biological and the psychosocial treatment approaches had benefits in the same direction, and disentangling the relative proportion of change accounted for by either one or both of the two technologies so far eluded the historian. There is no doubt that drug treatments played a significant role in changes in mental health care occurring during the 1950s, but exactly how much, whether it was exclusive, and how it interacted with the changes in social attitudes and the application of new psychosocial interventions is to be debated.

Although some professional and ideological tension existed—and still persists—between psychopharmacologists and social psychiatrists, in clinical practice the two technologies are readily combined, and the interaction between biological and social processes is recognized. The net result of combining these complementary approaches has been to greatly improve the outlook for the treatment and care of acute episodes. This process has had effects beyond the needs of individual patients. Professional confidence has increased, public attitudes have changed, the care of the mentally ill has been rapidly absorbed into the general health-care system, and the public monopoly on mental-health services has been altered and replaced by an emerging pluralistic, diversified system.

The carefully controlled research studies undertaken to establish the feasibility, efficacy, safety, and cost efficiency of these new treatment approaches are worthy of special note. In the early 1960s, a number of very ingenious research projects were undertaken to investigate the combination of drug and psychosocial treatments. These studies demonstrated, for example, that daycare treatment is an effective alternative to inpatient care for many patients. One notable treatment study was conducted by Zwerling and Wilder (1962) at Albert Einstein Medical School and indicated that as many as 80 percent of adult patients usually destined for inpatient hospital treatment could be treated in a day treatment program.

The studies also demonstrated that home treatment by nurses was another viable alternative to hospitalization. Pasamanick, Scarpitti, and Dinitz (1967) conducted a systematic investigation in Louisville and described in their important book, *Schizophrenics in the Community,* how intensive home treatment by visiting nurses combined with medication could prevent hospitalization, improve family functioning, and effect rapid symptom reduction.

It also has been shown that brief hospitalization is feasible and effective. The Veterans' Administration study was of crucial public health policy importance, as were the studies carried out at the Massachusetts Mental Health Center and Harvard Medical School by Greenblatt et al. (1965) in the Drugs and Social Therapies Project. Recent studies at the Langley Porter Neuropsychiatric Institute of the University of California at San Francisco (Glick et al. 1976) and at Columbia University (Herz et al. 1975, 1971) have confirmed the value of brief, intensive hospitalization for acute patients, including schizophrenics.

Social science research played a significant role throughout the 1950s and 1960s in the evolution of the mental-health-service system. Social scientists, represented most influentially by Goffman (1961), and journalists of that era were key figures in exposing the adverse and dehumanizing effects of large public institutions. During these two decades, a number of professionals such as Thomas Szasz and R.D. Laing and other outspoken civil libertarians criticized mental hospitals as being both ineffective and unjust, that is, institutions deprived patients not only of their health but also of their civil rights.

Sociologists and others invoked the term *institutionalism* to suggest that disordered behaviors were artifacts of the social institution and not manifestations of intrinsic psychiatric disturbances. Views were expounded that, at their most extreme, represented a romantic view that no entity such as schizophrenia existed. One of the hypotheses widely discussed was whether or not the most bizarre features of schizophrenia—hallucinations, social outbursts, and impaired thinking—might in fact be due not to an intrinsic disorder but to the institutional setting. These hypotheses implied that at some point the behaviors of institutionalized patients were responses to the total institutional environment rather than manifestations of the disorder.

In addition to the critics stressing the adverse effects of institutionalization were those stressing its social injustice. These critics asserted that mental hospitals were "total institutions," doomed to inescapable failure because of their special relationship to the larger society—that is, their status as "double agents" (their dual commitment to social control and personal change). The critics argued that when conflicts arose between the two goals of social control and treatment, the social-control mandate usually assumed priority over treatment. Because public mental hospitals are under the legal aegis, administrative control, and fiscal dominion of the larger society, they must be responsive to legislatures, commissions, and agencies who have as their highest priority the control of deviance rather than the needs of individual patients. Even innovations such as combining drugs and psychosocial methods were criticized as palliative and as a means to induce further conformity and social control of deviance.

The remedy the radical critics called for was deinstitutionalization. Decarceration became the battle cry of the new abolitionists. The radical reformers proposed the abolition of total institutions, particularly those employing involuntary commitment and involuntary treatment, and suggested that they be replaced by alternative forms of voluntary and community treatment for the mentally ill.

By the mid–1960s, supporters of deinstitutionalization policies including the social psychiatric reformers, the radical critics, and the civil libertarians were joined by the conservative budget administrators, who, pressured by strong fiscal forces, were eager to shift the burden of financing from the state to the federal level. While their theoretical rationales differed dramatically, these diverse groups formed a short-lived consensus on mental-health-care policies.

Federal efforts to reform the delivery of mental health services were initiated in the mid–1950s, when Congress established the Joint Commission on Mental Illness and Health. The resulting report and its recommendations led to the landmark 1963 address by President Kennedy to Congress on mental illness and retardation. Federal efforts at reform culminated in 1963 with the passage of the Mental Retardation Facilities and Community Mental Health Centers Construction Act, which established guidelines and promised federal funding for community facilities to treat mentally ill patients. This legislation marked a dramatic shift in treatment ideology that contributed to the radical changes in the mental-health-service system.

Whatever may have been their respective historical contributions, the result of new biological and psychosocial treatments and altered federal policies has been a radical change in the locus and structure of the mental-health-service system. The most publicized change has been the dramatic decline in the U.S. mental-hospital-resident census from a peak in 1955 of 559,000 persons to less than 200,000 persons in 1975, representing a 65 percent decrease (see figure 12–1).

Where have these patients gone? Approximately one-third of them have been transferred from one type of institution—the public mental hospital—into another—the nursing home. The passage of Medicare and Medicaid legislation in 1965 accelerated this process; some patients have merely been transferred from the roster of one type of facility to another as part of the change in funding mechanisms. However, that relocation process alone cannot account for the total decrease in census of public mental hospitals.

As a result of the forces evolving the 1950s, a large number of patients from public mental institutions have been released into the community. It is very difficult to find an accurate estimate of the numbers of these people, but their upper limit is approximately 750,000 to 1,000,000 persons.

Just as there are critics of institutionalization, so too are there critics of deinstitutionalization. Almost every major city in the United States has had in the recent past (and still has) exposes about the plight of ex-mental patients returned to the community, the difficulties arising from negative

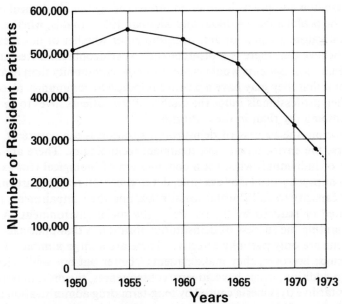

Figure 12-1. Number of Resident Patients at Year End in State and County
Mental Hospitals

community reactions, and questions raised regarding the humanity and
social utility of dumping patients into the community.

While the critics are quick to expose cases of the adverse consequences
of deinstitutionalization, the denominator for the statistics cited is unknown. Whereas it may be true that in a given city such as New York, there
are 70,000 discharged patients living in rooming houses and single-room
dwellings, a judgment as to the "failure" of deinstitutionalization cannot
be made without knowing what percentage of the total of discharge patients
this figure represents. It might be that the 70,000 persons represent a partial
adverse outcome of a total of 500,000 persons in New York who have been
discharged into the community, the majority of whom are functioning at an
acceptable level.

Other critics feel that the technologies—whether pharmacological or
psychosocial—that originally facilitated the shift from institutionalization
to community have become so extreme that they represent a form of behavior control for patients that is unacceptable on libertarian grounds (Klerman 1975).

In addition to concerns on civil-libertarian grounds, there is widespread
concern about the quality of care and quality of life of patients released
from institutions. The results of research to assess these variables are equiv-

ocal. Research and practice are often influenced by the ideological orientation of the professional (Armour and Klerman 1968), that is, whether or not the investigators are pro- or anti-institutional or whether or not they do or do not believe that there is a disorder such as schizophrenia. Research has shown that most patients would rather be in the community than in an institution and that they may have a positive response to community placement, even when professionals judge the quality of the patients' lives to be poorer in the community than in the institution.

Affecting the success of deinstitutionalization is the nature of mental disorders, the environment, and treatment technologies. There are a large number of individuals who, for a combination of biological (for example, genetic) or psychosocial (for example, developmental) reasons, are impaired in their capacity to fulfill usual social roles, and their impairments may be augmented or reduced by the nature of the social environment. A major factor limiting the success of deinstitutionalization is that pharmacological treatments are only partially effective. There are a large number of efficacious drugs; however, they make patients "better but not well" (Klerman 1977). In addition, drugs have adverse consequences, the most important of which is tardive dyskinesia caused by long-term drug administration to treat schizophrenia. Although the new technologies have helped greatly, particularly by reducing the bizarre, antisocial manifestations of schizophrenia and by maintaining patients without incarceration, a substantial proportion of these patients remain sufficiently handicapped, are unable to work, and are dependent on others for residential and financial assistance.

Restructuring of the Mental-Health-Service System

The President's Commission on Mental Health (1978) highlighted the need for relating the data on the use of mental-health services with the data on the prevalence of mental disorders. Psychiatric epidemiology, a relatively new branch of mental-health research, investigates how mental disorders are distributed in the population. Epidemiological investigations show the frequency with which various groups in the population (identified by age, ethnic background, education, urbanization of the place of residence, and social class) seek psychiatric treatment and the kind of treatment they seek. Advances in psychiatric epidemiology are resulting from innovative research strategies using improved measurement techniques with greater reliability and validity and are being influenced by recent scientific developments in diagnosis, genetics, psychopharmacology, neurobiology, and particularly psychopathology (Robins 1978; Weissman and Klerman 1978).

Regier and his colleagues (1978) summarized available epidemiological data and recent mental-health-services research findings to estimate the percent of the population with a mental disorder and the proportion utilizing various types of specialty mental-health and general-medical treatment settings. They concluded that at least 15 percent of the U.S. population is affected by some mental disorder in one year. In addition to that 15 percent (about 32,000,000 persons in 1975), there is a substantial percent, probably another 15 percent, who have symptoms of anxiety, stress, and distress related to problems of living. Many of these persons use the mental-health or general-health-care system for coping and consolation.

It is also possible to describe the specific disorders that comprise the 15 percent prevalence of mental disorders (see figure 12-2). Three disorders account for the predominance of this 15 percent prevalence: depression and affective disorders for 8 to 10 percent; alcoholism and related problems for up to 10 percent; and anxiety states for up to 7 percent. There is some overlap, especially between alcoholism and depression (Pottenger et al. 1978). Schizophrenia, the disorder that results in the greatest degree of institutionalization and that consumes the most public resources and attention due to its pervasiveness and high social disability, has a prevalence of about 1 percent. Another significant finding (Regier et al. 1978) is that the majority of the 15 percent of persons with mental disorders are in the health-care system. In 1975, however, only one-fifth of them were served in the specialty mental-health-care sector, three-fifths were identified in the general-medical-care sector, and one-fifth remained untreated.

The specialty mental-health-service system has undergone radical qualitative changes in its internal structure; these are reflected by a number of important quantitative changes. The total number of episodes treated in the mental-health-service system (that is, psychiatric hospitals, clinics, office practices) has increased dramatically from 1.7 million episodes in 1955 to

Disorders by Age Category	Point Prevalence (rate per 100 persons)
Children (under 18)	8-10%
Adults (18-65)	10-15%
Depression and Affective Disorders	4.5-8%
Anxiety, Phobia, and Other Neuroses	4-7%
Alcoholism and Alcohol Problems	2.5-8%
Drug Dependence	0.5-1%
Schizophrenia	0.5-1%
Aged (over 65)	10%

Figure 12-2. Estimated Prevalence of Selected Alcohol, Drug, and Mental Disorders

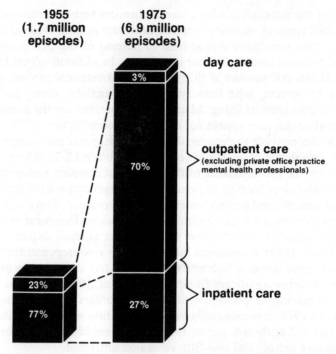

Figure 12–3. Percentage Distribution of Patient-Care Episodes in Mental-Health Facilities by Modality: United States, 1955 and 1975

6.9 million episodes in 1975 (see figure 12–3); almost the whole increase can be attributed to expansion in outpatient care. During this twenty-year period, there has been a proliferation of all types of outpatient facilities. In 1955, inpatient services accounted for 77 percent of the total number of episodes; but by 1975 this percentage decreased to 27 percent while outpatient episodes increased from 23 to 70 percent during the same period. When the data in absolute numbers are converted to rates (patient-care episodes per 100,000 population), it can be seen that rates for inpatient treatment have remained fairly constant at 800 episodes per 100,000 of population. In contrast, utilization of outpatient services has increased enormously.

Hospitalization, a component of the specialty mental-health-service system, has been the focus of most attention and concern because it is the most expensive part of the mental-health-care system. During the period from 1955 to 1975, there has also been a radical restructuring of the inpatient delivery system (see figure 12–4), although there has been relatively little growth in the hospitalization rate per 100,000 of population. However,

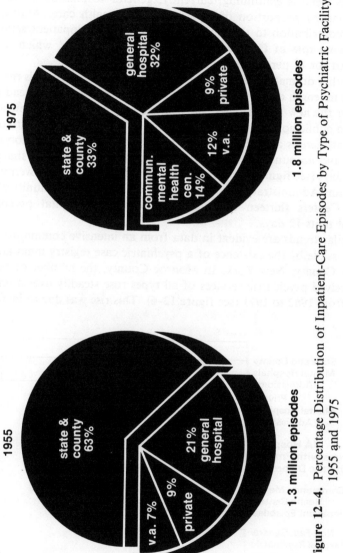

1.3 million episodes 1.8 million episodes

Figure 12-4. Percentage Distribution of Inpatient-Care Episodes by Type of Psychiatric Facility, 1955 and 1975

during this twenty-year interval the proportion of inpatient care accounted for by state and county hospitals decreased from 63 percent in 1955 to 33 percent in 1975, while there was a major increase in care in general hospitals. This trend is continuing; currently, general hospitalization accounts for the largest proportion of inpatient mental-health care. Another important contribution to the changing distribution of inpatient services is the growing role of Community Mental Health Centers, which in 1975 accounted for 14 percent of inpatient episodes.

Another dramatic change in the character of inpatient care is the sharp decrease in length of hospitalization. In 1954, before the widespread application of the new technologies, the average length of stay for new admissions in a public hospital was over six months. In 1975 median length of stay (note that there is a small percentage at one end of the distribution that is chronic and accounts for long-term hospitalization) decreased sharply to only 26 days (see figure 12–5). In private mental hospitals, it is twenty days, in Veterans Administration hospitals eighteen days, in Community Mental Health Centers thirteen days, and in nonfederal-hospital psychiatric-inpatient units 12 days.

Similar trends are evident in data from an intensive community study made possible by the existence of a psychiatric case registry maintained in Monroe County, New York. In Monroe County, the number of persons who received psychiatric services of all types rose steadily over a ten-year period from 1962 to 1971 (see figure 12–6). This rise was due to increasing

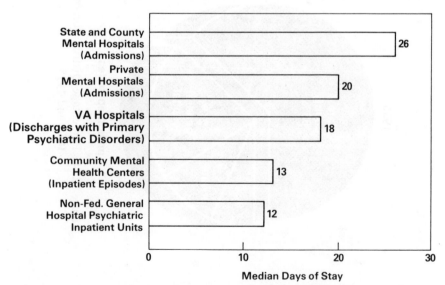

Figure 12–5. Length of Stay in Mental-Health Inpatient Settings, 1975

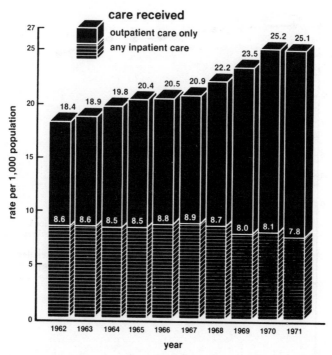

Figure 12-6. Annual Rate per 1000 Population of Persons Receiving
Psychiatric Services by Locus of Care: Monroe County,
New York, 1962-1971

utilization of outpatient services; utilization of inpatient care actually
declined. The average number of days, per person, receiving inpatient services also declined steadily over the ten-year period from 1962 to 1971 (see
figure 12-7).

These trends graphically depict the transformation of the mental-
health-service system during the past twenty-five years. This transformation is primarily the result of new forms of technology—both biological and
psychological—combined with changes in public attitudes and the fiscal
reimbursement system. Technological achievements have contributed to
reduced length of hospitalization, the control of psychoses, and the effectiveness of community care and treatment of many mental disorders.
Changes in public attitudes and reimbursement policies have resulted in a
major shift in the mental-health-service system from one with almost total
reliance on involuntary incarceration and treatment in public institutions to
a voluntaristic and pluralistic system. The shift to a voluntaristic system is
occurring as increasing numbers of patients have the choice of applying for

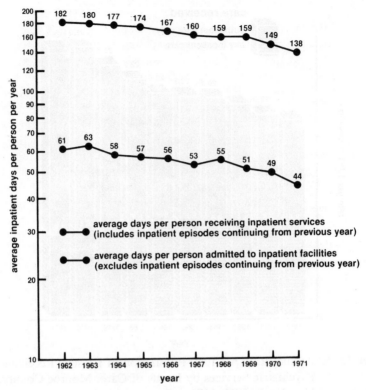

Figure 12-7. Average Number of Inpatient Days per Person per Year: Monroe County, New York, 1962–1971

treatment via their health insurance, and the system is becoming increasingly pluralistic as patients have more options regarding the types of treatment they will receive.

The Use of Minor Tranquilizers

Another consequence of the new drug technology, one that is the subject of a great deal of controversy, is the widespread use of antianxiety drugs, represented almost entirely by diazepam (Valium) and chlordiazepoxide (Librium). The three most prescribed drugs in the world as well as in the United States are in the following rank order: (1) Valium; (2) Darvon; and (3) Librium.

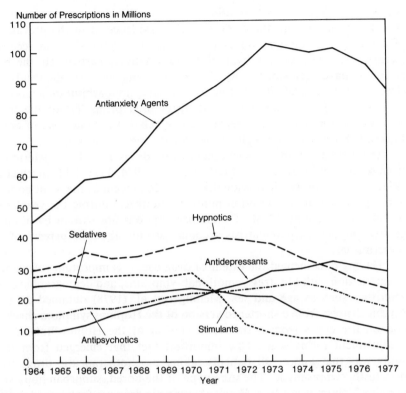

Number of Prescriptions in Millions

Sources: Balter 1974 and unpublished data.

Figure 12-8. Psychotherapeutic Drugs: Prescriptions Filled in U.S. Drugstores, 1964–1977

Changes between 1964 and 1977 in the absolute number of prescriptions for psychotherapeutic drugs are seen in figure 12–8. The number of prescriptions for antianxiety drugs far exceeds those for other classes of psychoactive drugs. The most striking trend, and the one that has become the subject of most discussions, is the enormous rise in the number of prescriptions for antianxiety drugs, which increased rapidly from 1964 to a peak of over 100,000,000 in 1973. Although this index of use has plateaued and even decreased slightly to 90,000,000 prescriptions in 1977, the level remains high by any criterion.

There have been some significant changes in other classes of psychotherapeutic drugs, but of relatively less magnitude. The use of stimulants, mainly amphetamines, has decreased dramatically since the early 1970s, mainly due to constraints on prescriptions caused by the awareness of their adverse effects and evidence for their limited efficacy in the treatment of obesity or of depression. The antidepressants have risen very slowly, not at all commensurate with the epidemiological evidence as to the prevalence of depression or the amount of public attention depression has received recently. Prescriptions for hypnotics and sedatives, particularly the barbiturates, have declined, mainly because of public concern about adverse effects and the availability of effective and safer nonbarbiturate hypnotics. Prescriptions for antipsychotics, the drugs used primarily for the treatment of schizophrenia, have remained at a relatively stable level of about 15,000.000 per year after a slight increase during the early 1970s.

These data regarding psychotherapeutic drugs have been determined from several national surveys (Mellinger et al. 1974, 1978; Uhlenhuth et al. 1974, 1978), all of which indicate that about 20 percent of the adult population reported using prescription minor tranquilizers during the year prior to being interviewed. Most of these prescriptions are written by nonpsychiatric physicians; psychiatrists account for only about 15 percent of the prescriptions.

Having documented the extent of drug use, it is important to examine its interaction with life stress, emotional symptoms, and demographic characteristics such as age and sex. Mellinger et al. (1978) obtained data on psychic distress from a shortened version of the Hopkins symptom checklist and on life crises from a shortened version of the Holmes-Rahe social readjustment rating scale. One important fact that emerged from their study was that almost all patients taking psychotherapeutic drugs were manifestly symptomatic. The stereotype of the bored, suburban housewife popping Valium is not true. Psychotherapeutic drug use (comprised mostly of minor tranquilizers) was clearly and strongly related to level of psychic distress and life crisis. For example, in the high-distress group, 30 percent had used psychotherapeutic drugs during the past year as compared with 8 percent of those classified as low on the distress measure.

Another way to interpret these figures is to note that 30 percent of the highly distressed persons used psychotherapeutic drugs whereas 70 percent did not. This 70 percent may be interpreted as untreated prevalence. Opponents of psychotherapeutic drug use see the 30 percent usage rate as too high whereas advocates see it as too low.

Mellinger et al. (1978) noted a striking sex difference in psychotherapeutic drug use: Women use these drugs about twice as frequently as men. One determinant of this difference may be that women go to physicians more than men so they are more likely to be prescribed drugs. Another determinant has to do with differences between men and women in reporting their symptoms. At all specified levels of life stress as measured by the Holmes-Rahe technique used in the Mellinger et al. study, women reported more symptoms than men. When combined ratings on psychic distress and

Table 12-1

Relation of Sex and Age to Psychotherapeutic Drug Use and Moderate-to-Heavy Alcohol Use in High-Distress Subjects

| Age Groups (Years) | Psychotherapeutic Drug Use/% | | | | Alcohol Use/% | | No. | |
| | Any Past Year | | Regular | | | | | |
	Women	Men	Women	Men	Women	Men	Women	Men
18–29	27	7	10	—	42	85	77	54
30–44	36	21	15	12	29	65	147	65
45–59	33	32	13	15	21	43	152	51
60–74	44	40	20	17	8	27	128	40

Source: Adapted from Mellinger et al. 1978.

life crisis were calculated, the disparity between the sexes in amount of psychotherapeutic drug use increased.

A complex interaction exists between selected independent variables—sex, age, and psychic distress—and the dependent variables of psychotherapeutic drug and alcohol use (see table 12-1). Psychotherapeutic drug use was found to increase with age in both men (more dramatically) and women with high distress, but exactly the reverse was true for alcohol. As age increased, the sex difference in drug use was diminished, but men of all ages used far more alcohol than did women.

Summary

In the interval of a mere quarter century, our country has witnessed the introduction of psychotherapeutic drugs, their use as potent agents to ameliorate severe mental disorders, and their widespread use to relieve symptoms of distress. The consequences have been both far-reaching and controversial. Drug therapies have transformed the entire mental-health-service system, enabling it to shift its locus from the institution to the community and causing a radical restructuring of facilities. Drug therapies are now also widely used in the general-health-care system, where they are frequently prescribed for relief of psychic distress. Although many points remain controversial, there is no doubt that new technologies and increased availability of mental-health services have resulted in improved treatment of mental disorders.

References

Armor, D.J., and Klerman, G.L. Psychiatric treatment orientations and professional ideology. *Journal of Health and Social Behavior* 9:243–255, 1968.

Caffey, E.M; Galbrect, C.R.; and Klett, C.J. Brief hospitalization and after care in the treatment of schizophrenia. *Archives of General Psychiatry* 24:81–86, 1971.

Glick, I.D.; Hargreaves, W.A.; Drues, J.; and Schostack, J.A. Short versus long hospitalization: A prospective controlled study. IV. One-year follow-up results for schizophrenic patients. *American Journal of Psychiatry* 133:509–514, 1976.

Goffman, E. *Asylums.* New York: Doubleday.

Greenblatt, M.; Solomon, M.; Evans, A.S.; and Brooks, G.W. *Drugs and Social Therapy in Chronic Schizophrenia.* Springfield: Ill.: Charles C Thomas, 1965.

Gruenberg, E.M. Evaluating the effectiveness of community mental health services. *Milbank Memorial Fund Quarterly,* Part II, 44, January, 1966.

Herz, M.I.; Endicott, J.; and Spitzer, R.L. Brief hospitalization of patients with families: Initial results. *American Journal of Psychiatry* 132:413–418, 1975.

Herz, M.I.; Endicott, J.; Spitzer, R.L.; and Mesnikoff, A. Day versus inpatient hospitalization: A controlled study. *American Journal of Psychiatry* 127:1371–1382, 1971.

Jones, M. *The Therapeutic Community.* New York: Basic Books, 1953.

Klerman, G.L. Behavioral control and the limits of reform. *Hastings Center Report* 5:40–45, 1975.

———. Better but not well: Social and ethical issues in the deinstitutionalization of the mentally ill. *Schizophrenia Bulletin* 3:617–31, 1977.

———. Catecholamine research: A paradigm for the relationship between basic investigations and clinical applications in psychiatry. In Usdin, E.; Kopin, I.J.; and Barchas, J. (eds.) *Catecholamines: Basic and Clinical Frontiers,* vol. 1, Proceedings of the Fourth International Catecholamine Symposium. New York: Pergamon Press, 1978, pp. 1–3.

Levine, J.; Schiele, B.; and Bouthilet, L. *Principles and Problems in Establishing the Efficacy of Psychotropic Agents.* Washington, D.C.: U.S. Government Printing Office, PHS Publication No. 2138, 1971.

Mellinger, G.D.; Balter, M.B.; Manheimer, D.I.; et al. Psychic distress, life crisis, and use of psychotherapeutic medications. *Archives of General Psychiatry* 35:1045–1052, 1978.

Mellinger, G.D.; Balter, M.B.; Parry, H.J.; et al. An overview of psycho-therapeutic drug use in the United States. In Josephson, E., and Carrol, E.E. *Drug Use: Epidemiological and Sociological Approaches.* Washington, D.C.: Hemisphere Publishing Corporation, 1974, pp. 333–366.

Pasmanick, B.; Scarpitti, F.R.; and Dinitz, S. *Schizophrenics in the Community: An Experimental Study in the Prevention of Hospitalization.* New York: Appleton-Century-Crofts, 1967.

Pottenger, M.; McKernon, J.; Patrie, L.E.; Weissman, M.M.; Ruben, H.L.; and Newberry, P. The frequency and persistence of depressive symptoms in the alcohol abuser. *Journal of Nervous and Mental Diseases* 166:562–570, 1978.

Regier, D.; Goldberg, I.D.; and Taube, C.A. The de facto U.S. mental health services system: A public health perspective. *Archives of General Psychiatry* 35:685–693, 1978.

Robins, L. Psychiatric epidemiology. *Archives of General Psychiatry* 35: 697–702, 1978.

Report to the President's Commission on Mental Health. Vol. I. Washington, D.C.: U.S. Government Printing Office, 1978.

Rosenthal, R., and Rosnow, R.L. *Artifact in Behavioral Research.* New York: Academic Press, 1969.

Swazey, J.P. *Chlorpromazine in Psychiatry.* Cambridge, Mass.: MIT Press, 1974.

Uhlenhuth, E.H.; Balter, M.B.; and Lipman, R.S. Minor tranquilizers clinical correlates of use in an urban population. *Archives of General Psychiatry* 31:759–764, 1978.

Uhlenhuth, E.H.; Lipman, R.S.; Balter, M.B.; and Stern, M. Symptom intensity and life stress in the city. *Archives of General Psychiatry* 31: 759–764., 1974.

Weissman, M.M., and Klerman, G.L. Epidemiology of mental disorders. *Archives of General Psychiatry* 35:705–712, 1978.

Zwerling, I., and Wilder, J.F. Day hospital treatment of psychotic patients. *Current Psychiatric Therapy* 2:200–210, 1962.

13 Ergotism: The Satan Loosed in Salem?

Linnda R. Caporael

Numerous hypotheses have been devised to explain the occurrence of the Salem witchcraft trials in 1692, yet a sense of bewilderment and doubt pervades most of the historical perspectives on the subject. The physical afflictions of the accusing girls and the imagery of the testimony offered at the trials seem to defy rational explanation. A large portion of the testimony, therefore, is dismissed as imaginary in foundation. One avenue of understanding that has yet to be sufficiently explored is that a physiological condition, unrecognized at the time, may have been a factor in the Salem incident. Assuming that the content of the court records is basically an honest account of the deponents' experiences, the evidence suggests that convulsive ergotism, a disorder resulting from the ingestion of grain contaminated with ergot, may have initiated the witchcraft delusion.

Suggestions of physical origins of the afflicted girls' behavior have been dismissed without research into the matter. In looking back, the complexity of the psychological and social factors in the community obscured the potential existence of physical pathology, suffered not only by the afflicted children, but also by a number of other community members. The value of such an explanation, however, is clear. Winfield S. Nevins best reveals the implicit uncertainties of contemporary historians [1;2, p. 235].

> . . . I must confess to a measure of doubt as to the moving causes in this terrible tragedy. It seems impossible to believe in tithe of the statements which were made at the trials. And yet it is equally difficult to say that nine out of every ten of the men, women and children who testified upon their oaths, intentionally and wilfully falsified. Nor does it seem possible that they did, or could invent all these marvelous tales; fictions rivaling the imaginative genius of Haggard or Jules Verne.

The possibility of a physiological condition fitting the known circumstances and events would provide a comprehensible framework for understanding the witchcraft delusion in Salem.

Originally published as L. Caporael "Ergotism, The Satan Loosed in Salem?" *Science* 192 (2 April 1976):21–26. Copyright © 1976 by the American Association for the Advancement of Science. Reprinted with permission.

Background

Prior to the Salem witchcraft trials, only five executions on the charge of witchcraft are known to have occurred in Massachusetts [3, 4]. Such trials were held periodically, but the outcomes generally favored the accused. In 1652, a man charged with witchcraft was convicted of simply having told a lie and was fined. Another man, who confessed to talking with the devil, was given counsel and dismissed by the court because of the inconsistencies in his testimony. A bad reputation in the community combined with the accusation of witchcraft did not necessarily insure conviction. The case against John Godfrey of Andover, a notorious character consistently involved in litigation, was dismissed. In fact, soon after the proceedings, Godfrey sued his accusers for defamation and slander and won the case.

The supposed witchcraft at Salem Village was not initially identified as such. In late December 1691, about eight girls, including the niece and daughter of the minister, Samuel Parris, were afflicted with unknown "distempers" [4-6]. Their behavior was characterized by disorderly speech, odd postures and gestures, and convulsive fits [7]. Physicians called in to examine the girls could find no explanation to their illness, and in February one doctor suggested the girls might be bewitched. Parris seemed loath to accept this explanation at the time and resorted to private fasting and prayer. At a meeting at Parris's home, ministers from neighboring parishes advised him to "sit still and wait upon the Providence of God to see what time might discover" [6, p. 25].

A neighbor, however, took it upon herself to direct Parris's Barbados slave, Tituba, in the concocting of a "witch cake" in order to determine if witchcraft was present. Shortly thereafter, the girls made an accusation of witchcraft against Tituba and two elderly women of general ill repute in Salem Village, Sarah Good and Sarah Osborn. The three women were taken into custody on 29 February 1692. The afflictions of the girls did not cease, and in March they accused Martha Corey and Rebecca Nurse. Both of these women were well respected in the village and were convenanting members of the church. Further accusations by the children followed.

Examinations of the accused were conducted in Salem Village until 11 April by two magistrates from Salem Town. At that time, the examinations were moved from the outlying farming area to the town and were heard by Deputy Governor Danforth and six of the ablest magistrates in the colony, including Samuel Sewall. This council had no authority to try accused witches, however, because the colony had no legal government—a state of affairs that had existed for 2 years. By the time Sir William Phips, the new governor, arrived from England with the charter establishing the government of Massachusetts Bay Colony, the jails as far away from Salem as Boston were crowded with prisoners from Salem awaiting trial. Phips appointed a special Court of Oyer and Terminer, which heard its first case on 2 June. The proceedings resulted in conviction, and the first condemned witch was hanged on 10 June.

Before the next sitting of the court, clergymen in the Boston area were consulted for their opinion on the issues pending. In an answer composed by Cotton Mather, the ministers advised "critical and exquisite caution" and wished "that there may be as little as possible of such noise, company and openness as may too hastily expose them that are examined" [2, p. 83]. The ministers also concluded that spectral evidence (the appearance of the accused's apparition to an accuser) and the test of touch (the sudden cessation of a fit after being touched by the accused witch) were insufficient evidence for proof of witchcraft.

The court seemed insensitive to the advice of the ministers, and the trials and executions in Salem continued. By 22 September, 19 men and women had been sent to the gallows, and one, Giles Corey, had been pressed to death, an ordeal calculated to force him to enter a plea to the court so that he could be tried. The evidence used to obtain the convictions was the test of touch and spectral evidence. The afflicted girls were present at the examinations and trials, often creating such pandemonium that the proceedings were interrupted. The accused witches were, for the most part, persons of good reputation in the community; one was even a former minister in the village. Several notable individuals were "cried out" upon, including John Alden and Lady Phips. All the men and women who were hanged had consistently maintained their innocence; not one confessor to the crime was executed. It had become obvious early in the course of the proceedings that those who confessed would not be executed.

On 17 September 1692, the Court of Oyer and Terminer adjourned the witchcraft trials until 2 November; however, it never met again to try that crime. In January 1693 the Superior Court of Judicature, consisting of the magistrates on the Court of Oyer and Terminer, met. Of 50 indictments handed in to the Superior Court by the grand jury, 20 persons were brought to trial. Three were condemned but never executed and the rest were acquitted. In May Governor Phips ordered a general reprieve, and about 150 accused witches were released. The end of the witchcraft crisis was singularly abrupt [2, 4, 8].

Tituba and the Origin Tradition

Repeated attempts to place the occurrences at Salem within a consistent framework have failed. Outright fraud, political factionalism, Freudian psychodynamics, sensation seeking, clinical hysteria, even the existence of witchcraft itself, have been proposed as explanatory devices. The problem is primarily one of complexity. No single explanation can ever account for the delusion; an interaction of them all must be assumed. Combinations of interpretations, however, seem insufficient without some reasonable justification for the initially afflicted girls' behavior. No mental derangement or fraud seems adequate in understanding how eight girls, raised in soul-searching Puritan tradition, simultaneously exhibited the same symptoms or conspired together for widespread notoriety.

All modern accounts of the beginnings of Salem witchcraft begin with Parris's Barbados slave, Tituba. The tradition is that she instructed the minister's daughter and niece, as well as some other girls in the neighborhood, in magic tricks and incantations at secret meetings held in the parsonage kitchen [2, 4, 8, 9]. The odd behavior of the girls, whether real or fraudulent, was a consequence of these experiments.

The basis for the tradition seems twofold. In a warning against divination, John Hale wrote in 1702 that he was informed that one afflicted girl had tried to see the future with an egg and glass and subsequently was followed by a "diabolical molestation" and died [6]. The egg and glass (an improvised crystal ball) was an English method of divination. Hale gives no indication that Tituba was involved, or for that matter, that a group of girls was involved. I have been unable to locate any reference that any of the afflicted girls died prior to Hale's publication.

The other basis for the tradition implicating Tituba seems to be simply the fact that she was from the West Indies. The Puritans believed the American Indians worshiped the devil, most often described as a black man [4]. Curiously, however, Tituba was not questioned at her examination about activities as a witch in her birthplace. Historians seem bewitched themselves by fantasies of voodoo and black magic in the tropics, and the unfounded supposition that Tituba would inevitably be familiar with malefic arts of the Caribbean has survived.

Calef [7] reports that Tituba's confession was obtained under duress. She at first denied knowing the devil and suggested the girls were possessed. Although Tituba ultimately became quite voluble, her confession was rather pedestrian in comparison with the other testimony offered at the examinations and trials. There is no element of West Indian magic, and her descriptions of the black man, the hairy imp, and witches flying through the sky on sticks reflect an elementary acquaintance with the common English superstitions of the time [9–11].

Current Interpretations

Fraud

Various interpretations of the girls' behavior diverge after the discussion of its origins. The currently accepted view is that the children's symptoms of affliction were fraudulent [4, 8, 12]. The girls may have perpetrated fraud simply to gain notoriety or to protect themselves from punishment by adults as their magic experiments became the topic of rumor [2]. One author supposes that the accusing girls craved "Dionysiac mysteries" and that some were "no more seriously possessed than a pack of bobby-soxers on the loose" [8, p. 29]. The major difficulty in accepting the explanation of purposeful fraud is the gravity of the girls' symptoms; all the eyewitness accounts agree to the severity of the affliction [6, 10, 11, 13].

Upham [4] appears to accept the contemporaneous descriptions and ascribes to the afflicted children the skills of a sophisticated necromancer. He proposes that they were able ventriloquists, highly accomplished actresses, and by "long practice" could "bring blood to the face, and send it back again" [4, vol. 2, p. 395]. These abilities and more, he assumes, the girls learned from Tituba. As discussed above, however, there is little evidence that Tituba had any practical knowledge of witchcraft. Most colonists, with the exception of some of the accused and their defenders, did not appear even to consider pretense as an explanation for the girls' behavior. The general conclusion of the New Englanders after the tragedy was that the girls suffered from demonic possession [2, 6, 9].

Hysteria

The advent of psychiatry provided new tools for describing and interpreting the events at Salem. The term hysteria has been used with varying degrees of license [2, 8, 9, 14], and the accounts of hysteria always begin in the kitchen with Tituba practicing magic. Starkey [8] uses the term in the loosest sense; the girls were hysterical, that is overexcited, and committed sensational fraud in a community that subsequently fell ill to "mass hysteria." Hansen [9] proposes the use of the word in a stricter, clinical sense of being mentally ill. He insists that witchcraft really was practiced in Salem and that several of the executed were practicing witches. The girls' symptoms were psychogenic, occasioned by guilt at practicing fortune-telling at their secret meetings. He states that the mental illness was catching and that the witnesses and majority of the confessors became hysterics as a consequence of their fear of witchcraft. However, if the girls were not practicing divination, and if they did indeed develop true hysteria, then they must all have developed hysteria simultaneously—hardly a credible supposition. Furthermore, previous witchcraft accusations in other Puritan communities in New England had never brought on mass hysteria.

Psychiatric disorder is used in a slightly different sense in the argument that the witchcraft crisis was a consequence of two party (pro-Parris and anti-Parris) factionalism in Salem Village [4]. In this account, the girls are unimportant factors in the entire incident. Their behavior "served as a kind of Rorschach test into which adults read their own concerns and expectations" [14, p. 30]. The difficulty with linking factionalism to the witch trials is that supporters of Parris were also prosecuted while some nonsupporters were among the most vociferous accusers [2, 14]. Thus, it becomes necessary to resort to projection, transference, individual psychoanalysis, and numerous psychiatric disorders to explain the behavior of the adults in the community who were using the afflicted children as pawns to resolve their own personal and political differences.

Of course, there was fraud and mental illness at Salem. The records clearly indicate both. Some depositions are simply fanciful renditions of local gossip or cases of malice aforethought. There is also testimony based on exaggerations of nightmares and inebriated adventures. However, not all the records are thus accountable.

Physiological Explanations

The possibility that the girls' behavior had a physiological basis has rarely arisen, although the villagers themselves first proposed physical illness as an explanation. Before accusations of witchcraft began, Parris called in a number of physicians [6, 7]. In an early history of the colony, Thomas Hutchinson wrote that "there are a great number of persons who are willing to suppose the accusers to have been under bodily disorders which affected their imagination" [12, vol. 2, p. 47]. A modern historian reports a journalist's suggestion that Tituba had been dosing the girls with preparations of jimson-weed, a poisonous plant brought to New England from the West Indies in the early 1600's [8, footnote on p. 284]. However, because the Puritans identified no physiological cause, later historians have failed to investigate such a possibility.

Ergot

Interest in ergot (*Claviceps purpura*) was generated by epidemics of ergotism that periodically occurred in Europe. Only a few years before the Salem witchcraft trials the first medical scientific report on ergot was made [15]. Denis Dodart reported the relation between ergotized rye and bread poisoning in a letter to the French Royal Académie des Sciences in 1676. John Ray's mention of ergot in 1677 was the first in English. There is no reference to ergot in the United States before an 1807 letter by Dr. John Stearns recommending powdered ergot sclerotia to a medical colleague as a therapeutic agent in childbirth. Stearns is generally credited with the "discovery" of ergot; certainly his use prompted scientific research on the substance. Until the mid-19th century, however, ergot was not known as a parasitic fungus, but was thought to be sunbaked kernels of grains [15-17].

Ergot grows on a large variety of cereal grains—especially rye—in a slightly curved, fusiform shape with sclerotia replacing individual grains on the host plant. The sclerotia contain a large number of potent pharmacologic agents, the ergot alkaloids. One of the most powerful is isoergine (lysergic acid amide). This alkaloid, with 10 percent of the activity of D-LSD (lysergic acid diethylamide) is also found in ololiuqui (morning glory seeds), the ritual hallucinogenic drug used by the Aztecs [15, 16].

Warm, damp, rainy springs and summers favor ergot infestation. Summer rye is more prone to the development of the sclerotia than winter rye, and one field may be heavily ergotized while the adjacent field is not. The fungus may dangerously parasitize a crop one year and not reappear again for many years. Contamination of the grain may occur in varying concentrations. Modern agriculturalists advise farmers not to feed their cattle grain containing more than one to three sclerotia per thousand kernels of grain, since ergot has deleterious effects on cattle as well as on humans [16, 18].

Ergotism, or long-term ergot poisoning, was once a common condition resulting from eating contaminated rye bread. In some epidemics it appears that females were more liable to the disease than males [19]. Children and pregnant women are most likely to be affected by the condition, and individual susceptibility varies widely. It takes 2 years for ergot in powdered form to reach 50 percent deterioration, and the effects are cumulative [18, 20]. There are two types of ergotism—gangrenous and convulsive. As the name implies, gangrenous ergotism is characterized by dry gangrene of the extremities followed by the falling away of the affected portions of the body. The condition occurred in epidemic proportions in the Middle Ages and was known by a number of names, including *ignis sacer,* the holy fire.

Convulsive ergotism is characterized by a number of symptoms. These include crawling sensations in the skin, tingling in the fingers, vertigo, tinnitus aurium, headaches, disturbances in sensation, hallucination, painful muscular contractions leading to epileptiform convulsions, vomiting, and diarrhea [16, 18, 21]. The involuntary muscular fibers such as the myocardium and gastric and intestinal muscular coat are stimulated. There are mental disturbances such as mania, melancholia, psychosis, and delirium. All of these symptoms are alluded to in the Salem witchcraft records.

Evidence for Ergotism in Salem

It is one thing to suggest convulsive ergot poisoning as an initiating factor in the witchcraft episode, and quite another to generate convincing evidence that it is more than a mere possibility. A jigsaw of details pertinent to growing conditions, the timing of events in Salem, and symptomology must fit together to create a reasonable case. From these details, a picture emerges of a community stricken with an unrecognized physiological disorder affecting their minds as well as their bodies.

Growing Conditions

The common grass along the Atlantic Coast from Virginia to Newfoundland was and is wild rye, a host plant for ergot. Early colonists were dissatisfied with it as forage for their cattle and reported that it often made the cattle ill with unknown diseases [22]. Presumably, then, ergot grew in the New World before the Puritans arrived. The potential source for infection was already present, regardless of the possibility that it was imported with English rye.

Rye was the most reliable of the Old World grains [22] and by the 1640s it was a well-established New England crop. Spring sowing was the rule; the bitter winters made fall sowing less successful. Seed time for the rye was April and the harvesting took place in August [23]. However, the grain was stored in barns and often waited months before being threshed when the weather turned cold. The timing of Salem events fits this cycle. Threshing probably occurred shortly before Thanksgiving, the only holiday the Puritans observed. The children's symptoms appeared in December 1691. Late the next fall, 1692, the witchcraft crisis ended abruptly and there is no further mention of the girls or anyone else in Salem being afflicted [4, 9].

To some degree or another all rye was probably infected with ergot. It is a matter of the extent of infection and the period of time over which the ergot is consumed rather than the mere existence of ergot that determines the potential for ergotism. In his 1807 letter written from upstate New York, Sterns [15, p. 274] advised his medical colleague that, "On examining a granary where rye is stored, you will be able to procure a sufficient quantity [of ergot sclerotia] from among that grain." Agricultural practice had not advanced, even by Sterns's time, to widespread use of methods to clean or eliminate the fungus from the rye crop. In all probability, the infestation of the 1691 summer rye crop was fairly light; not everyone in the village or even in the same families showed symptoms.

Certain climatic conditions, that is, warm, rainy springs and summers, promote heavier than usual fungus infestation. The pattern of the weather in 1691 and 1692 is apparent from brief comments in Samuel Sewall's diary [24]. Early rains and warm weather in the spring progressed to a hot and stormy summer in 1691. There was a drought the next year, 1692, thus no contamination of the grain that year would be expected.

Localization

"Rye," continues Sterns [15, p. 274], "which grows in low, wet ground yields [ergot] in greatest abundance." Now, one of the most notorious of the accusing children in Salem was Thomas Putnam's 12-year-old daughter, Ann. Her mother also displayed symptoms of the affliction and psychological historians have credited the senior Ann with attempting to resolve her own neurotic complaints through her daughter [8, 9, 14]. Two other afflicted girls also lived in the Putnam residence. Putnam had inherited one of the largest landholdings in the village. His father's will indicates that a large measure of the land, which was located in the western sector of Salem Village, consisted of swampy meadows [25] that were valued farmland to the colonists [22]. Accordingly, the Putnam farm, and more broadly, the western acreage of Salem Village, may have been an area of contamination. This contention is further substantiated by the pattern of residence of the accusers, the accused, and the defenders of the accused living within the boundaries of Salem Village (figure 12-1). Excluding the afflicted girls, 30 of 32 adult accusers lived in the western section and 12 of the 14 accused witches lived in the eastern section, as did 24 of the 29 defenders [14]. The general pattern of residence, in combination with the well-documented factionalism of the eastern and western sectors, contributed to the progress of the witchcraft crisis.

The initially afflicted girls show a slightly different residence pattern. Careful examination reveals plausible explanations for contamination in six of the eight cases.

Three of the girls, as mentioned above, lived in the Putnam residence. If this were the source of ergotism, their exposure to ergotized grain would be natural. Two afflicted girls, the daughter and niece of Samuel Parris, lived in the parsonage almost exactly in the center of the village. Their exposure to contaminated grain from western land is also explicable. Two-thirds of Parris's salary was paid in provisions; the villagers were taxed proportionately to their landholding [4]. Since Putnam was one of the largest landholders and an avid supporter of Parris in the minister's community disagreements, an ample store of ergotized grain would be anticipated in Parris's larder. Putnam was also Parris's closest neighbor with afflicted children in residence.

The three remaining afflicted girls lived outside the village boundaries to the east. One, Elizabeth Hubbard, was a servant in the home of Dr. Griggs. It seems plausible that the doctor, like Parris, had Putnam grain, since Griggs was a professional man, not a farmer. As the only doctor in town, he probably had many occasions to treat Ann Putnam, Sr., a woman

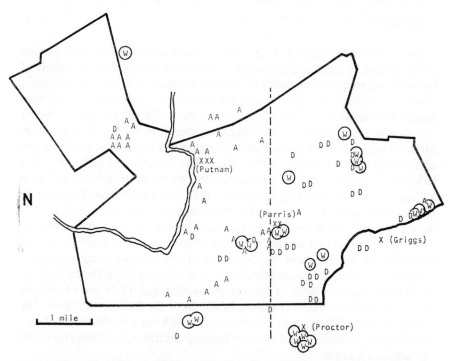

Source: Adapted from Boyer and Nissenbaum [14, 25]. The names in parentheses indicate the households in which the afflicted girls were living, exluding Sarah Churchill, whose affliction is believed to have been fraudulent. The nonvillagers shown on this map are those whose places of residence lay on the fringes of the Village boundaries. Residences are labeled X, afflicted girl; W, accused witch; D, defender of the accused; and A, accuser.

Figure 13-1. Residence Patterns, Salem Village, 1692

known to have much ill health [2, 4]. Griggs may have traded his services for provisions or bought food from the Putnams.

Another of the afflicted, Sarah Churchill, was a servant in the house of a well-off farmer [25]. The farm lay along the Wooleston River and may have offered good growing conditions for ergot. It seems probable, however, that Sarah's affliction was a fraud. She did not become involved in the witchcraft persecutions until May, several months after the other girls were afflicted, and she testified in only two cases, the first against her master. One deponent claimed that Sarah later admitted to belying herself and others [11].

How Mary Warren, a servant in the Proctor household, would gain access to grain contaminated with ergot is something of a mystery. Proctor had a substantial farm to the southeast of Salem and would have had no need to buy or trade for food. Both he and his wife were accused of witchcraft and condemned. None of the Proctor children showed any sign of the affliction; in fact, three were accused and imprisoned. One document offered as evidence against Proctor indicates that Mary stayed overnight in the village [11]. How often she stayed or with whom is unknown.

Mary's role in the trials is particularly curious. She began as an afflicted person, was accused of witchcraft by the other afflicted girls, and then became afflicted again. Two depositions filed against her strongly suggest, however, that at least her first affliction may have been a consequence of ergot poisoning. Four witnesses attested that she believed she had been "distempered" and during the time of her affliction had thought she had seen numerous apparitions. However, when Mary was well again, she could not say that she had seen any specters [11]. Her second affliction may have been the result of intense pressure during her examination for witchcraft crimes.

Ergotism and the Testimony

The utmost caution is necessary in assessing the physical and mental states of people dead for hundreds of years. Only the sketchiest accounts of their lives remain in public records. In the case of ergot, a substance that affects mental as well as physical states, recognition of the social atmosphere of Salem in early spring 1692 is basic to understanding the directions the crisis took. The Puritans' belief in witchcraft was a totally accepted part of their religious tenets. The malicious workings of Satan and his cohorts were just as real to the early colonists as their belief in God. Yet, the low incidence of witchcraft trials in New England prior to 1692 suggests that the Puritans did not always resort to accusations of black magic to deal with irreconcilable differences or inexplicable events.

The afflicted girls' behavior seemed to be no secret in early spring. Apparently it was the great consternation that some villagers felt that induced Mary Sibley to direct the making of the witch cake of rye meal and the urine of the afflicted. This concoction was fed to a dog, ostensibly in the belief that the dog's subsequent behavior would indicate the action of any malefic magic [14]. The fate of the dog is unknown; it is quite plausible that it did have convulsions, indicating to the observers that there was witchcraft involved in the girls' afflictions. Thus, the experiments with the witch cake, rather than any magic tricks by Tituba, initiated succeeding events.

The importance of the witch cake incident has generally been overlooked. Parris's denouncement of his neighbor's action is recorded in his church records. He clearly stated that, until the making of the cake, there was no suspicion of witchcraft and no reports of torturing apparitions [4]. Once a community member had gone "to the Devil for help against the Devil" as Parris put it, the climate for the trials had been established. The afflicted girls, who had made no previous mention of witchcraft, seized upon a cause for their behavior—as did the rest of the community. The girls named three persons as witches and their afflictions thereby became a matter for the legal authorities rather than the medical authorities or the families of the girls.

The trial records indicate numerous interruptions during the proceedings. Outbursts by the afflicted girls describing the activities of invisible specters and "familiars" (agents of the devil in animal form) in the meeting house were common. The girls were often stricken with violent fits that were attributed to torture by apparitions. The spectral evidence of the trials appears to be the hallucinogenic symptoms and perceptual disturbances accompanying ergotism. The convulsions appear to be epileptiform [6, 13].

Accusations of choking, pinching, pricking with pins, and biting by the specter of the accused formed the standard testimony of the afflicted in almost all the examinations and trials [26]. The choking suggests the involvement of the involuntary muscular fibers that is typical of ergot poisoning; the biting, pinching, and pricking may allude to the crawling and tingling sensations under the skin experienced by ergotism victims. Complaints of vomiting and "bowels almost pulled out" are common in the deposition of the accusers. The physical symptoms of the afflicted and many of the other accusers are those induced by convulsive ergot poisoning.

When examined in the light of a physiological hypothesis, the content of so-called delusional testimony, previously dismissed as imaginary by historians, can be reinterpreted as evidence of ergotism. After being choked and strangled by the apparition of a witch sitting on his chest, John Londer testified that a black thing came through the window and stood before his face. "The body of it looked like a monkey, only the feet were like cock's feet, with claws, and the face somewhat more like a man's than a monkey . . . the thing spoke to me . . ." [25, p. 45].

Joseph Bayley lived out of town in Newbury. According to Upham [4], the Bayleys, en route to Boston, probably spent the night at the Thomas Putnam residence. As the Bayleys left the village, they passed the Proctor house and Joseph reported receiving a "very hard blow" on the chest, but no one was near him. He saw the Proctors, who were imprisoned in Boston at the time, but his wife told him that she saw only a "little maid." He received another blow on the chest, so strong that he dismounted from his horse and subsequently saw a woman coming toward him. His wife told him she saw nothing. When he mounted his horse again, he saw only a cow where he had seen the woman. The rest of Bayley's trip was uneventful, but when he returned home, he was "pinched and nipped by something invisible for some time" [11]. It is a moot point, of course, what or how much Bayley ate at the Putnams', or that he even really stayed there. Nevertheless, the testimony suggests ergot. Bayley had the crawling sensations in the skin, disturbances in sensations, and muscular contractions symptomatic of ergotism. Apparently his wife had none of the symptoms and Bayley was quite candid in so reporting.

A brief but tantalizing bit of testimony comes from a man who experienced visions that he attributed to the evil eye cast on him by an accused witch. He reported seeing about a dozen "strange things" appear in his chimney in a dark room. They appeared to be something like jelly and quavered with a strange motion. Shortly, they disappeared and a light the size of a hand appeared in the chimney and quivered and shook with an upward motion [27]. As in Bayley's experience, this man's wife saw nothing. The testimony is strongly reminiscent of the undulating objects and lights reported in experiences induced by LSD [28].

By the time the witchcraft episode ended in the late fall 1692, 20 persons had been executed and at least two had died in prison. All the convictions were obtained on the basis of the controversial spectral evidence [2]. One of the commonly expressed observations about the Salem Village witchcraft episode is that it ended unexpectedly for no apparent reason [2, 4]. No new circumstances to cast spectral evidence in doubt occurred. Increase Mather's sermon on 3 October 1692, which urged more conclusive evidence than invisible apparitions or the test of touch, was just a stronger reiteration of the clergy's 15 June advice to the court [2]. The grounds for dismissing the spectral evidence had been consistently brought up by the accused and many of their defenders throughout the examinations. There had always been a strong undercurrent of opposition to the trials and the most vocal individuals were not always accused. In fact, there was virtually no support in the colonies for the trials, even from Boston, only 15 miles away. The most influential clergymen lent their support guardedly at best; most were opposed. The Salem witchcraft episode was an event localized in both time and space.

How far the ergotized grain may have been distributed is impossible to determine clearly. Salem Village was the source of Salem Town's food supply. It was in the town that the convictions and orders for executions were obtained. Maybe the thought processes of the magistrates, responsible and respected men in the Colony, were altered. In the following years, nearly all of them publicly admitted to errors of judgment [2]. These posttrial documents are as suggestive as the court proceedings.

In 1696, Samuel Sewall made a public acknowledgment of personal guilt because of the unsafe principles the court followed [2]. In a public apology, the 12 jurymen stated [9, p. 210], "We confess that we ourselves were not capable to understand nor able to withstand the mysterious delusion of the Powers of Darkness and Prince of the Air . . . [we] do hereby declare that we justly fear that we were sadly deluded and mistaken. . . ." John Hale, a minister involved in the trials from the beginning, wrote [6, p.

167], "such was the darkness of the day . . . that we walked in the clouds and could not see our way."

Finally, Ann Putnam, Jr., who testified in 21 cases, made a public confession in 1706 [2, p. 250].

> I justly fear I have been instrumental with others though ignorantly and unwittingly, to bring upon myself and this land the guilt of innocent blood; though what was said or done by me against any person I can truly and uprightly say before God and man, I did it not for any anger, malice or ill will to any person, for I had no such things against one of them, but what I did was ignorantly, being deluded by Satan.

One Satan in Salem may well have been convulsive ergotism.

Conclusion

One could reasonably ask whether, if ergot was implicated in Salem, it could have been implicated in other witchcraft incidents. The most cursory examination of Old World witchcraft suggests an affirmative answer. The district of Lorraine suffered outbreaks of both ergotism [15] and witchcraft persecutions [4] periodically throughout the Middle Ages until the 17th century. As late as the 1700s, the clergy of Saxony debated whether convulsive ergotism was symptomatic of disease or demonic possession [17]. Kittredge [3], an authority on English witchcraft, reports what he calls "a typical case" of the early 1600s. The malicious magic of Alice Trevisard, an accused witch, backfired and the witness reported that Alice's hands, fingers, and toes "rotted and consumed away." The sickness sounds suspiciously like gangrenous ergotism. Years later, in 1762, one family in a small English village was stricken with gangrenous ergotism. The Royal Society determined the diagnosis. The head of the family, however, attributed the condition to witchcraft because of the suddenness of the calamity [29].

Of course, there can never be hard proof for the presence of ergot in Salem, but a circumstantial case is demonstrable. The growing conditions and the pattern of agricultural practices fit the timing of the 1692 crisis. The physical manifestations of the condition are apparent from the trial records and contemporaneous documents. While the fact of perceptual distortions may have been generated by ergotism, other psychological and sociological factors are not thereby rendered irrelevant; rather, these factors gave substance and meaning to the symptoms. The content of hallucinations and other perceptual disturbances would have been greatly influenced by the state of mind, mood, and expectations of the individual [30]. Prior to the witch cake episode, there is no clue as to the nature of the girls' hallucinations. Afterward, however, a delusional system, based on witchcraft, was generated to explain the content of the sensory data [31, p. 137]. Valins and Nisbett [31, p. 141], in a discussion of delusional explanations of abnormal sensory data, write, "The intelligence of the particular patient determines the structural coherence and internal consistency of the explanation. The cultural experiences of the patient determine the content—political, religious, or scientific—of the explanation." Without knowledge of ergotism and confronted by convulsions, mental disturbances, and perceptual distortions, the New England Puritans seized upon witchcraft as the best explanation for the phenomena.

References and Notes

1. I have attempted to use sources that would be readily available to any reader. The spelling of quotations from old documents has been modernized to promote clarity.

2. W.S. Nevins, *Witchcraft in Salem Village* (Franklin, New York, 1916; reprinted 1971).

3. G.L. Kittredge. *Witchcraft in Old and New England* (Harvard Univ. Press, Cambridge, Mass., 1929).

4. C.W. Upham, *Salem Witchcraft* (Wiggins & Lunt, Boston, 1867; reprinted by Ungar, New York, 1959), vols. 1 and 2.

5. The number of afflicted girls varies between 8 and 12, depending on the history consulted. I have restricted the "afflicted girls" to those eight whose residence in or near Salem Village is known. They are Ann Putnam, Jr., Mary Warren, Mercy Lewis, Sarah Churchill, Betty Parris, Abigail Williams, Elizabeth Hubbard, and Mary Walcott.

6. J. Hale, *A Modest Inquiry Into the Nature of Witchcraft* (Boston: 1702; facsimile reproduction by York Mail, Bainbridge, N.Y., 1973).

7. R. Calef, in *Narratives of the Witchcraft Cases 1648–1706,* G.L. Burr, ed. (Scribner's, New York, 1914).

8. M.L. Starkey, *The Devil in Massachusetts* (Knopf, New York, 1950).

9. C. Hansen, *Witchcraft in Salem* (Braziller, New York, 1969).

10. S.G. Drake, *The Witchcraft Delusion in New England* (Franklin, New York, 1866; reprinted 1970).

11. W.E. Woodard, Records of Salem Witchcraft (privately printed, Roxbury, 1864; reprinted by Da Capo, New York, 1969).

12. T. Hutchinson, *The History of the Colony and Province of Massachusetts Bay,* L.S. Mayo, ed. (Harvard Univ. Press, Cambridge, Mass., 1936), vols. 1 and 2.

13. D. Lawson, in *Narratives of the Witchcraft Cases 1648–1706,* G.L. Burr, ed. (Scribner's, New York, 1914).

14. P. Boyer and S. Nissenbaum, *Salem Possessed: The Social Origins of Witchcraft* (Harvard Univ. Press, Cambridge, Mass., 1974) (a map indicating the geography of the witchcraft is on p. 35).

15. F.J. Bove, *The Story of Ergot* (Barger, New York, 1970).

16. A. Hoffer, *Clin. Pharmacol. Ther.* 6, 183 (1965).

17. G. Barger, *Ergot and Ergotism* (Gurney & Jackson, London, 1931).

18. C.E.Sajous and J.W. Hundley, *The Cyclopedia of Medicine* (Davis, Philadelphia, 1937), vol. 5, pp. 412–416.

19. Ergot has been used to induce and hasten labor in childbirth; however, it is generally unsuccessful in procuring abortion. Also, there is no evidence that epidemics of chronic convulsive ergotism of the type hypothesized to have occurred in Salem have produced abortions [17].

20. C.M. Gruber, *The Cyclopedia of Medicine, Surgery, Specialties* (Davis, Philadelphia, 1950). vol. 5, pp. 245–248.

21. W.C. Cutting, *Handbook of Pharmacology: Action and Uses of Drugs* (Appleton-Century-Crofts, New York, 1972).

22. L. Carrier, *The Beginnings of Agriculture in America* (McGraw-Hill, New York, 1923).

23. R.E. Walcott, *N. Engl. Q.* 9, 218 (1936).

24. M.H. Thomas, ed., *The Diary of Samuel Sewall 1674–1729* (Farrar, Straus & Giroux, New York, 1973).

25. P. Boyer and S. Nissenbaum, eds., *Salem Village Witchcraft: A Documentary Record of Local Conflict in Colonial New England* (Wadsworth, Belmont, Calif., 1972). The editors publish an extremely useful map adapted from Upham [4].

26. A random selection of almost any testimony in Woodard [11] will attest to this.

27. Essex County Archives, Salem Witchcraft. Elizer Keysar's testimony from the Thomas F. Madican photostats as transcribed in the Works Progress Administration verbatim report, vol. 2, p. 9.

28. S.H. Snyder, *Madness and the Brain* (McGraw-Hill, New York, 1974).

29. D. van Zwanenber, *Med. Hist. 17, 204 (1973).*

30. *A. Goth, Medical Pharmacology* (Mosby, St. Louis, 1972).

31. S. Valins and R. Nisbett, in *Attribution: Perceiving the Causes of Behavior,* E. Jones, et al., eds. (General Learning Press, Morristown, N.J., 1972).

32. I thank C.F. Paul and M.B. Brewer for their helpful comments on the manuscript.

14 The Social Control of Biology

G. Russell Carpenter

Darwin did not include himself in his own thought. True, he did see himself as having a body, a biological constitution, and an evolutionary foundation. But his thought, his acts of theorizing, are not discussed. He did not say much about the theory of evolution's effect on the course of evolution or of the effect of science on biology. Yet this is the question we face. Gene cloning, plasmid research, is life creating life. We are fast coming on a bio-technique—using life as a factory system—that is unparalleled in history. In trying to plan the future of this we will have all Darwin's silences to deal with—all the questions concerning thought in the scheme of things and the rational control of evolution.

There is a sense in which talking about the social control of biology is talking about too much. Obviously every civilization *is* the social control of biology. The domestication of food is herding the wild. The sexual contract is a manipulation of the gene pool. Then, too, there are effects of medical practices, wars, chemicals, pollution, and man-made radiation to consider. The relationship between human civilization and the evolution of species is becoming ever more complex. Indeed, one could even consider the space program's effect on biology. One could look at the way we attempt to domesticate the ecosphere with high technology and cybernetics. All of this represents too much to consider in a single model; the ecological relationship between symbolic cultures and biological foundation is just too involved. No single discourse, theory, or perspective is sufficient to the intellectual demands of it.

Nevertheless, ecological crises grow worldwide; more and more slow-to-die pesticides and poisonous chemicals show up in out-of-the-way places; sea-dumped chemicals, some radioactive, begin to leak into fisheries; there are ground sores of old chemicals in dumps, some under cities, some leaking into streams. And there are other familiars—carbon dioxide build-ups, the greenhouse effect and fluorocarbons, ozone depletion, acid rain, deforestation, overpopulation, and so on—all of which, in one way or another, represent the human control of biology. Though not always the kind of control we intended, all are the products of human thought and the forms of our civilization.

Although just recently recognized, these crises have been some time in

281

the making. In part, they emerge from the "epistemological thaw" that gave us modern thought. That is, they relate to the coming of instrumental rationality, to the triumph of reduction, to the victories of paradigmatic science, and to the separation of fact from value, and so they relate to the dangers inherent in a fragmenting scientific consciousness, and in applying specific remedies to specific problems. These crises (as opposed to those outside the compass of Western thought—famine, disease beyond the socio-political reach of microbiology) also relate to a change in the nature and deployment of social power. Michel Foucault (1978, p. 142) wrote:

> Western man was gradually learning what it meant to be a living species in a living world, to have a body, conditions of existence, probabilities of life, an individual and collective welfare, forces that could be modified. . . . For the first time in history, no doubt, biological existence was reflected in political existence.

Social power, Foucault argues, is today "situated and exercised" at the level of the life of a population. Biological existence is now waged in politics. Modern forms of power seek to establish a norm, a range of the normal in the population, and then to regenerate this norm again and again by fostering, supervising, regulating, and administering its existence and by retarding deviations—such as epidemics, overpopulations, delinquencies, and addictions. Foucault calls this form of power "a biopolitics of the population." "The old power of death that symbolized sovereign power was now carefully supplanted by the administration of bodies and the calculated management of life" (1978, p. 140). So at the very moment when scientists were arguing that their knowledge was objective, disinterested, and basic, science became entangled with politics. Inescapably, science provides politics with both the definition of reality (acceptable levels, known dangers, possible defenses) and with some of the conditions (nuclear threats, carcinogenic chemicals, mind-bending drugs) for political regulation. Still, this administering, this fostering and supervising of life through history, is anything but a unified practice. Not only is it partial and piecemeal, but oftentimes—as is now commonly known—one solution causes another problem. We are not close to a rational management of the biosphere, but then there is nothing else to expect from a specialized, huge, and fragmenting science selectively applied in bureaucratic political space.

I shall argue here that the new naturalism, this urge to connect a biological foundation to social phenomena, is an old urge and is in many ways a denial of the recursive nature of human praxis representing the exhausting of a form of the modern imagination. In doing so I shall argue that given modern ecological crises, more important than the way one's biology determines one's social existence is the way rationality *as* social existence influences the biosphere.

But even beyond that eschatological concern of mine lies a basic theoretical issue that troubles the new naturalism: If we draw the determinist's arrow from, say, genes to behavior, we cannot do so in all cases, most notably we cannot do it in that case where the behavior in question concerns gene biology itself, for here the very genetic foundation that determines, delimits, conditions, or influences thought may be changed by thought itself. This basic indeterminacy shadows every aspect of the urge to find a natural foundation for man and his societies. Nonetheless, without addressing the "nature" of rational transcendence—the naturalness of our sciences, if you will—pretentions of planetary management of biology are just that, pretentions. It is in the face of this latter that I argue the new naturalism is the exhausting of a form of our imagination, of little use in the search for solutions, and that it will have little to add to the social control of biology precisely because it avoids these issues. Something else is required.

In between Individual Thought, Species Rules

The urge to connect behavior to genes theoretically is the urge to search for the origins of things. In it a particular grounding is sought, but only in a particular way. Attempting to avoid the epistemological insecurities involved in studying the symbolic/social, the new naturalists attempt to step around the Hegelian hall of mirrors by promoting a variant of natural science. But this attempt to circumvent the issues can be no more theoretically justified than, say, the structuralist's or the cultural determinist's attempt to ground behavior in the foundation of its codes of culture or the materialist's attempt to do so in the means of production. All represent the same urge to find an explanatory grounding, one in the foundation of language, another in the workings of human production, and the other in genetic history.

But unlike the work of the new naturalists, the other two seem more actively stung by the problem of specifying the working of representation in their theories. For example, by bringing up the concept of false consciousness, Marxists always speak into the teeth of its threat. One can ask, how can they be sure they are not profoundly hoodwinked by a history more cunning than that history they think they know? That is, how can they be sure they are not falsely conscious? They cannot be sure. This is the reason their claimed transcendence is justified by eschatology, by the claim of the eventual redemption of belief. (The point is not to understand but to change, said Marx.) So too with the cultural determinist. By recognizing and addressing the relativity of cultural systems, anthropologists and sociologists necessarily bring up the problem of the relativity of their own culture and its unconscious operations.

With the recognition of relativity in cultures, the notion of progress becomes troubled; it is no longer easy to see the savage without seeing oneself. It gets harder to justify that there is a distance between the pen and the stick. Thus, the old dilemma: When symbolic studies of symbolics recognize that they are damned without recourse to the unconscious in culture, they begin to experience trouble justifying the special nature of their studies. So again the question arises: How do they know their view is better than the one they are studying? Who is to say anthropological representations are of a higher order than the savages'?

The new naturalists attempt to avoid facing the problem of symbolic representation by keeping to positivist forms of explanation and by hiding the problem of fragmentation in their theories, behind the concept of paradigm. Fragmentation, here, comes to be called *paradigmatic relativity;* this is an old social science ploy.

Then too, there is another more subtle manner in which they attempt to slip by the issue of representation: They selectively avoid questions concerning the genetic foundations of science, even though science is clearly social behavior and related to human genes. By not addressing this issue they let go of all the important concerns about knowledge and human interest. These are cast out of view. If man's chemistry is disturbing the biosphere, then it is the science and politics of chemistry that should be looked to, not the chemistry in man's DNA. Likewise with the hydrogen bomb, the question of the genetic necessity of nuclear war cannot be handled by the theories of the new naturalists, even though if there were such a war it would dramatically influence the gene pool and the biosphere. The reason this question cannot be addressed is the necessity of including science in the things theorized about.

A science looking at science faces the hall of mirrors and the problems of representation that destroyed philosophy and broke the social sciences. So these problems are avoided and this avoidance is justified on the basis of the very looseness in the genetic code. Francois Jacob (1973, p. 316) wrote:

> With the development of the nervous system, with learning and memory, the rigor of heredity is relaxed. In the genetic programme underlying the characteristics of a fairly complex organism, there is a closed part that can only be expressed in a fixed way and another part that allows the individual a certain freedom of response. . . . Here it commands; there it permits.

Obviously, the more the new naturalists claim they are interested in the command side, the more they can claim their topic phenomena are lawlike and their study, science. Then subtly: Obviously science is on the permit side; the important limits of inquiry are the shared rules of a paradigm, not those in the genetic code.

This subtle near-exempting of science from the topic domain allows the new naturalists to lean on the discursive arrangements the physical sciences use to warrant their own work *and* to limit the epistemological self-criticism concerning the presuppositions of science. But it does not allow the socio-biologists, the biosociologists, and whoever traffics in the natural bases of social phenomena to evade the problem of representation.

Allow me to use Foucault's version to illustrate. According to Foucault (1970), western man sees himself as the "enslaved sovereign." Clearly this too is the perspective of the new naturalists. For them, man is sovereign enough to theorize about his or her genetic enslavement, yet the naturalists' theories assert there are limits on genetics. Enslavement on the one hand and freedom on the other. But the point lies, of course, with the theory and with what it says, with the separaton of the two, with knowing which is which and when and how they work. One can always ask how we are sure of (1) the reality of the nonconscious things that we theorize are enslaving us, how we are sure they actually exist when we are sufficiently sovereign to imagine, and (2) how we know that the theorizing actually is done in the domain of conscious freedom. That is, in the case of those who would theo-retically connect genes to behavior, how are we sure DNA is not actually retaining the real determining conditions in the unconscious?

The modern intellectual task, Foucault asserts, is to explain what is unconscious in existence to what is conscious. The demand, he says, is to work "backward—or downwards—to an analytic of finitude, in which man's being will be able to provide a foundation in their own positivity for all those forms that indicate to him that he is not infinite" (1970, p. 315). But then this task requires two positions, one *empirical*—a particular spe-cies nature, a particular linguistic embeddedness, participation in a partic-ular form of the means of production—and the other *transcendental*—one somehow free from the constraints of biology and a particular language and position in economics. "Man," Foucault writes (1970, p. 318), "in the ana-lytic of finitude, is a strange empirico-transcendental doublet, since he is a being such that knowledge will be attained in him of what renders all knowl-edge possible." Because both are there, because along with empirical man there is always his unconscious nature (the history of his language, species, means of production), the doublet is unstable. We simply cannot be sure (from the inside) that the representations we make to ourselves concerning our unconscious nature are not subtly or even profoundly in error.

And this is true for the biological determinists. They posit themselves to be within their own theoretical conditions: Whatever are the biological rules of our species behavior, they are the rules of the biological determinists too. If these rules affect the manner in which theorists theorizing about uncon-scious biological determinism think, then they do so always, in Gunter Remmling's words, "in Hegelian fashion, 'behind our backs'" (1967, p. 5). Saying such theorizing is science and letting it go at that will not do here because the new naturalism lacks a theoretical rule for knowing that place or point in man's nature where the determinism in the genetic program gives way to the wild freedom of individual thought. Simply said, they cannot justify that science is always on the transcendental side of the doublet.

Nothing prevents one from employing the new naturalism's analytic to the social behavior of doing science, from seeing science as genetically determined—nothing save the fact that doing so would bring into question the nature of making these "scientific" representations of things. And that of course would bring into question the presuppositions of a science that rests its philosophy on the concept of paradigm; it would, in other words, bring into question the form of theorizing in the new naturalism. Hence comes the need to include only some behaviors in the topic domain and the demand to exclude others. The excluded ones, unfortunately (man's high technology, modern science, his instrumental rationality) seem the important ones now that our social power is situated at the level of the population.

Foucault (1978, p. 143) writes: "For millenia, man remained what he was for Aristotle: a living animal with the additional capacity for political existence; modern man is an animal whose politics places his existence as a living being in question." The important consideration, it would seem, lies with representing this latter condition to ourselves, not with explaining the animal side of our existence. The question is how natural is pollution, and what can we do about that?

In between Management and the Unintended

We work literally up against the limits in the forms of our imagination. Where classical thought divided nature into human nature and the rest of nature and then assumed that human nature by the act of naming the rest could know it, we find we cannot do this. We have an epistemological awareness of man. Foucault again (1970, p. 317):

> Where there had formerly been a correlation between a *metaphysics* of representation and of the infinite and an *analysis* of living beings, of man's desires, and of the words of his language, we find being constituted an *analytic* of finitude and human existence, and in opposition to it (though in correlative opposition) a perpetual tendency to constitute a *metaphysics* of life, labor, and language.

"A perpetual tendency to constitute a metaphysics." Surely this describes Marxist thought, as well as the modern philological directions in the studies of culture and the attempt to create a new naturalism. Still, the operational term is *tendency*. We never succeed.

> Modern thought will contest even its own metaphysical impulses, and show that reflections upon life, labour, and language, in so far as they have value as analytics of finitude, express the end of metaphysics; the philosophy of

life denounces metaphysics as a veil of illusion, that of labour denounces it as an alienated form of thought and an ideology, that of language as a cultural episode. (Foucault 1970, p. 317)

The end of metaphysics is the negative side. The positive side is the appearance of man, this being who is the "observed spectator," this object of our studies. The problem is that we can locate no epistemologically clean position from which to view ourselves, from which to view the nature of our relation to nature. The science of ecology that attempts to study man's relationship to the environment is always incomplete because it cannot, finally, include its own work in the scheme of things. That is, ecology cannot include the effect of the science of ecology on ecology. That seems to require a metatheoretical position that we cannot find. The tendency, as Foucault says, is to then attempt one anyway. The impulse is to state certain conditions as historical absolutes and then reify those into the foundation for the rest.

The new naturalists attempt to find that foundation in the nature of the genetic code, but this does not represent a closing of social science concerns with those of biology. Left hanging are all the questions concerning the place of Darwin in his own theories, those questions about the relationships there are between the cultural and the natural, those questions about the coevolution of thought and the unthought. Those questions still seem beyond our representations and theories.

The review of evidence that should concern us here ought to be with the evidence of our failures to think our way through. If our genes matter, they seem to matter only in the final instance. Trying to make a science of the influence of genes when the products of science pollute the planet or threaten if not species annihilation then certainly an unprecedented biological disaster seems beside the point. Trying to make a science of that when modern man is rapidly coming upon the ability to manipulate our genetic endowment rationally seems not the urge to understand fully the workings of the genetic program. In other words: from blue-green algae, the genetic code works its way to a point where it thinks, then to where it thinks about itself, then to where it manipulates itself by thinking. This is the important evidence to consider. If we have a basic nature, it is this.

References

Foucault, Michel. 1970. *The Order of Things*. New York: Pantheon.
———. 1978. *The History of Sexuality*. New York: Pantheon.
Jacob, Francois. 1973. *The Logic of Life*. New York: Pantheon.
Remmling, Gunter W. 1967. *Road to Suspicion*. New York: Appleton-Century-Crofts.

15

A Restatement of the Issue

Walter R. Gove and
G. Russell Carpenter

In the introductory chapter we argued that sociology, as well as its sister disciplines, is in dire straits. It was demonstrated that there is a common theme in the literature that sociology qua sociology is not moving forward, but, instead, if anything, is becoming fragmented. It is the first author's personal sense that, although advances have been made in methodology and excellent studies conducted on particular subjects in the period since he entered graduate school in 1960, the discipline of sociology as a cumulative body of knowledge had not advanced and that there are increasing signs of stagnation. If this is true, or even partially true, then it would seem that we need a new way to approach the study of human behavior. In particular, we need a new way of conceptualizing our subject matter. As was suggested in chapter 1, the most critical issue is to stop treating nature and nurture as separate phenomena and to treat the biological, the psychological, and the social as phenomena inevitably linked to each other and as necessary components to be included in the study of human behavior.[1]

This book contains a number of empirical demonstrations that biological, psychological, and social phenomena interact in complex ways. It has been shown that the state and behavior of the individual are affected by biological and societal phenomena (chapters 5, 6, 7, 8, 9, 11, 12, and 13), as is diagrammed in figure 15-1. Similarly, it has been shown that the state and behavior of the individual result in change in biological and social phenomena (chapters 4, 5, 12, and 14), as is diagrammed in figure 15-2. And finally, it has been shown that social phenomena affect biological phenomena and vice versa (chapters 4, 5, 6, 10, 12, and 14), as is shown in figure 15-3.

The effect of biological phenomena on social phenomena has, of course, been one of the consistent themes in anthropology. This is particularly true with regard to age or sex differences and in terms of characteristics of the habitat and how this affects how food is produced and how this in turn affects the social structure. However, now we see that the issue is much more complicated than the interaction presented by anthropologists.

The first component of this conclusion was written primarily by Walter R. Gove and the second component was written primarily by G. Russell Carpenter.

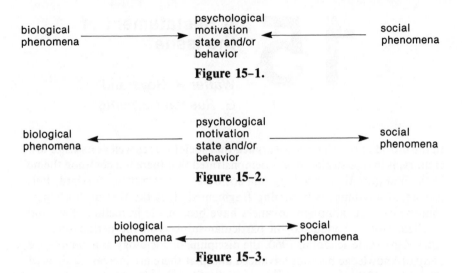

Figure 15-1.

Figure 15-2.

Figure 15-3.

As Klerman and Schlecter (chapter 12) have shown, the development of chemical agents (tranquilizers and antidepressants) has had an effect on the biological state of persons who are mentally ill, which in turn has profound effects on their psychological state and on their behavior. These effects have resulted, at the societal level, in a drastic restructuring with regard to how the mentally ill are treated, and, at the personal level, totally changed the career of those who are mentally ill. Furthermore, our society routinely produces physical products that yield undesirable biologial effects, such as various carcinogens and radiation. In short, we live in a physical environment that contains socially produced matter that has profound effects on our behavior and health. To make matters even more complicated, we are now launched into an era of genetic engineering in which we can produce patented biological organisms that are going to have a substantial effect on our lives.

In summary, the basic model of the determinants of our state and behavior are presented in figure 15-4, which shows that biological, psychological, and social phenomena all interact in a recursive system with the directions of causal relationships all being strong. This being the case, we are always in an ongoing interacting system composed of biological and social factors, and there is no beginning point and no way to step outside the system. As a consequence, the epistemological issues are extremely complex (chapter 2).

As noted above, the various empirical chapters provide evidence for figure 15-4. These chapters do not apply the conceptual or methodological breakthrough that appears to be called for if sociologists and other

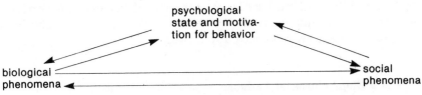

Figure 15-4.

investigators of human behavior are going to launch their fields or disciplines in an effective or profitable fashion. They are, in fact, examples of normal science that point to the *need* for a reconceptualization and structuring of how we view and study human behavior. The chapters reflect what is occurring on the periphery of a number of disciplines, and what is unique about this book is that it has brought together examples of the diversity of evidence that demonstrates the need for a reorientation. Cortez's chapter on the causes of crime and delinquency comes the closest to providing a new model for thinking about and analyzing the causes of particular forms of human behavior. His theoretical orientation (which has not been well received), along with Weiner's, comes close to providing an orientation of how one should proceed. But the orientation by itself does not provide the conceptual or methodological breakthroughs that are necessary.

When we organized the conference that led to this book, we felt the need for these conceptual or methodological breakthroughs. Unfortunately, we did not, and do not, have any to offer. Hence, the purpose of this conference and this book has been to demonstrate the need for a reorientation that would eventually produce such breakthroughs. Such an orientation in principle probably calls for the development of a behavioral science (not sciences) and perhaps a withering away of sociology and its sister disciplines as distinct disciplines with a clearly staked-out domain of study. However, if we are correct in our assessment, and the present endeavor is indicative of the future, such changes undoubtedly will come slowly, with the structural reorientation of the discipline studying human behavior following, and not preceeding, successful endeavors at reconceptualizing and studying human behavior.

Many of the chapters provide something of a model of how one should orient the way in which one *studies* a particular dependent variable. Equally important is the way one *conceptualizes* one's variables. Age is an excellent variable to consider. We think that one should use age as an explanatory variable in looking at a variety of human behaviors, and we will discuss how it might be used to understand certain aspects of deviant behavior. An inverse relationship of age to a diverse set of deviant behaviors is widely known, but the relationship has been largely ignored because various investigators and theoreticians have lacked a theoretical framework for explaining the relationships observed. We would argue that a biopsychosocial perspective of age provides such a framework.

Age as a Biopsychosocial Variable

In the area of the sociology of deviance, age is by far the most powerful predictor of the cessation of those forms of deviant behavior that involve substantial risk and/or a demanding life-style. This is true of most forms of crime, particularly those involving violence, theft, or high risk, whether measured by official statistics (Uniform Crime Reports 1971) or by self-reports (Rowe and Tittle 1977). Rates of alcoholism drop sharply with age (Robbins 1975), as do rates of drug addiction (McAuliffe 1975) and suicide attempts (Jarvis et al. 1976). Similarly, data in a national probability sample show both the simple use and onset of use of tobacco, alcohol, marijuana, and stimulants have a strong relationship with age (Crutchfield 1980; Mellinger et al. 1978). Likewise, surveys of mental health show poor mental health to be very strongly related to being young (Brocki 1979; Gaitz and Scott 1972). The evidence suggests that extramarital sex is limited primarily to the young (Booth and Edwards 1977). Similarly, persons with an antisocial personality (also referred to as sociopaths or psychopaths) tend to lose the characteristics of an antisocial personality with age (American Psychiatric Association 1980).

Although age is often used as a control variable in various exploratory studies, it rarely is used as an explanatory variable even though the relationship with age may be emphasized (for instance, Brocki 1979; Gaitz and Scott 1972; Robbins 1975). Furthermore, to the best of our knowledge, the few efforts to determine why age is related to the cessation of various forms of deviance have been unsuccessful. Rowe and Tittle (1977, p. 234), for example, state, "We readily admit that a really satisfying account of the age/crime relationship has eluded us." They go on to state, "we offer the problem as a challenge to other researchers because we believe that an answer to this question is critical for explaining crime."

In our view, if the social, psychological, and biological aspects associated with age were simultaneously considered, then it would be intuitively obvious why age has a negative relationship with this diverse form of deviant behavior. Society structures the life course of the individual by providing the stages which that individual sequentially goes through (Foner and Kertzer 1978; Gordon 1972). This appears to be a common characteristic across societies, although there is a sharp difference in the rigidity of the stages and in the extent to which these stages are marked by sharp transition points. Furthermore, in virtually all societies, the later stages are generally preferable to the early stages, with the possible exception of the last stage (Foner and Kertzer 1978).

One of the characteristics of any society—but one that is particularly a characteristic of a relatively fluid, complex, industrial society such as our own—is that a person has to discover the particular complex set of niches one is going to occupy. When one is in high school, the individual is uncertain as to whether he or she will marry, whom he or she will marry, when he or she will marry, and whether he or she will have children. One is uncertain about the processes of schooling (potentially up through graduate school or professional school) and about getting launched into a job or career, and how successful that job or career will be. In short the process of finding out who one is going to be is a process of waiting, effort, and chance; it is a process that involves feedback, both painful and pleasant, that results in adaptive behavior and changing expectations; and this increase in skills explains the fact that their measure of satisfaction increases with age. Campbell et al. (1976) argue that through this process one develops increasing skills both with regard to learning how to perform and with regard to knowing what to expect. The argument is essentially that persons find their niches with the passage of time and experience, and that realistic expectations are associated with increased levels of satisfactions and a muting of feelings of elation and despair. In their words (Campbell et al. 1976, p. 157):

> On balance we suspect that there are factors associated with the passage of time and the process of aging that have a major, if not exclusive, hand in generating the characteristic form of the age-satisfaction relationship. In other words, older people in general have a marked tendency to consider themselves more satisfied with their experience in most domains than younger people do, unless there is a clear and urgent set of influences to the contrary, as there obviously is where declining health is concerned.

Campbell et al. review evidence from their study as well as from the work of others, particularly Helson (1964), and conclude that there seem to be "important psychological mechanisms operative such that beyond a certain initial point of familiarity, *satisfaction with a situation increases as one becomes increasingly accommodated to it*" (italics in original). Brim (1968) and Pressey and Kuhlen (1957) take the view that as one finds one's niche in society, one becomes increasingly satisfied with one's situation.

In summary, from a sociological perspective, aging suggests that (at least until very old age) age statuses become increasingly comfortable with one's niche in society. This, of course, runs counter to the image that youth is the preferred age in our society. However, the evidence in support of this view is fairly impressive. At the anecdotal level we have Neugarten's (1968, p. 97) respondent who said, "There is a difference between wanting to feel young and wanting to be young. Of course, it would be pleasant to maintain the vigor and appearance of youth; but I would not trade those things for the authority or autonomy I feel—no, nor for the ease of interpersonal relationships nor the self-confidence that comes from experience." At the quantitative level we have the national probability surveys analyzed by Campbell et al. (1976) and Brocki (1979), who found a slight increase with life satisfaction with aging, Brocki also finding an increase of self-esteem.

At the psychological level there is an extensive literature on life stages that focuses first on the psychological problems involved in coming to find and establish one's place in society, coming to care for others and then to find a maturational stage where one comes to terms with one's life (see, Jung 1963, 1964, 1968; Erickson 1963, 1968; Lidz 1968; Gordon 1972; Sears and Feldman 1964; Levinson et al. 1978; Lowenthal et al. 1975). Virtually all researchers working in this area see persons shifting from self-absorption to caring for others and developing wisdom and maturity. To a very large extent the focal concerns highlighted by these diverse theoreticians paralleled the writings by sociologists who, as noted above, tend to see sequential age categories as becoming more desirable and to see the discovery-development and adaptation to one's niche as leading to increasing satisfaction.

A component of human behavior that is difficult to describe is what we will call *psychic energy*. By psychic energy we mean one's mental motivation, drive and willingness to persist in an activity despite various forms of hardship. By psychic energy we do not mean the strength of one's attitudes or convictions, but whatever it is that motivates us and calls us into mental and physical action. It seems very clear that after the age of 30 years or so most persons experience a drop-off in the amount of psychic energy. Less widely recognized, but well established by the work of a number of investigators (Back and Gergen 1966; Cumming and Henry 1961; Dean 1962; Gaitz and Scott 1972; Kuhlen 1959; Campbell 1976; Brocki 1979), this decline is found with both negative and positive measures on individual affective or mood states. Thus, with age persons experience less anguish, despair, and distress; they also experience less elation. In short, it appears that psychic energy declines with age. Why this is the case is unclear, but we would note that psychic energy appears to be closely tied to physical energy, which also appears to decline with age. The flattening of affect is also what one would expect given the discussions of the life course presented above.

One of the most obvious and readily documented aspects of age is a clear decline in one's physical abilities once one has passed the age of 30 years (Finch and Heylick 1973; Sears and Feldman 1964; Spirluso 1980), and that this is paralleled by a decline in one's physical energy.

The Age-Deviance Link

The forms of deviance noted above tend to reflect self-centered and socially problematic behaviors that require high levels of physical energy, and such energy declines with age (for example, it is physically harder to work when one is hung over and thus, as one ages, one is increasingly unwilling to place oneself in the situation where one has to work and is hung over). Such behaviors are also inconsistent with the social niche and being satisfied with one's life. They are also inconsistent with the psychological variables discussed, namely, the focal concerns that develop with aging, and the loss of psychic energy associated with aging. And finally, it is inconsistent with a deteriorating physical body or a decline in physical health. In short, if we consider the biopsychosocial aspects of aging, we will find that these three aspects all contain features that help explain the negative relationship between age and a large set of deviant behaviors.

As we have conceptualized the issue of the relationship between age and many forms of deviant behavior, we must simultaneously treat age as reflecting an individual's physical energy, physical health, psychic energy, and his or her social niche and focal concerns. Although there will be a strong tendency for these components of age to change in a consistent pattern, the rate of change across these components will vary from individual to individual. To study empirically the link between age and deviant behavior, we need to be in a situation where we can study the various components of aging. However, we do not have a clear conception of what is meant by either physical energy of psychic energy, to say nothing of measuring these variables. The components of physical health that need to be measured are unclear. The degree to which an individual is lodged in a particular social (and usually, complex) niche has never been studied directly. Similarly, the literature on focal concerns has been largely theoretical, and how to properly measure an individual's particular focal concerns remains problematic. In short, not only does the study of the relationship between age and deviant behavior involve biological, psychological, and social phenomena, but with regard to the particular variable to be studied, we at present have only a crude conceptualization of the variables and are not in a position to adequately measure them. In summary, taking a biopsychosocial approach to that relationship, it is easy to understand intuitively why the relationship occurs. But at the same time, we are not equipped to study the relationship and develop a clear understanding of intricacies of the pattern of the relationship. At the present time, the issue is that of conceptualization and measurement and has relatively little to do with the issue of statistical or modeling techniques.

Stepping Back: Where Are We?

The fragmentation of social thought more than anything else is responsible for the conference and this book. Somewhere in the debate the two of us have had for years—one of us an empiricist in a world of too many facts, the other a theorist in a world of too many theories—there was agreement that the center had not held; we were already past the point of return; the social sciences were deteriorating.

We feel that the easy use of the concept of paradigm is defensive. Social scientists use it as a shield when confronted with the relativity and fragmentation of modern social thought, but they do so emphatically, as though to quash any further discussion of the factual nature of the assertion (or of the implicit assumption) that the social sciences are on the other side of the epistemological threshold of science. The use of the concept allows practitioners to tell themselves they are doing "normal science" and to go on doing it. But this just continues the crisis we are in. Descriptive studies proliferate; survey-based (thus, historical) articles take over the journals, yet they connect to one another only in the rigor of a chain of citations; fundamental theoretical and philosophical issues are neglected in the reduction of these into problems of research methodology, and then, those into problems of technique. The result, we feel, is not a coming, but a lived, intellectual sterility in the social sciences.

The modern intellectual task, Michel Foucault (1970) contends, is to explain that which is unconscious in existence to that which is conscious. If that is true, then these discussions, eclectic and diverse as they are, represent nothing but the expression of that basic urge in social thought. And, unfortunately, they all suffer from the contradictions implicit in the task of working backward or downward toward a foundation that is already there and working in the unconscious. In an analytical, though not necessarily the same eschatological, sense our tasks here are little different than those of Marx or Freud. Freud said, "Where the id was, the ego shall be," and all of his work—the construction of a mythology of desire, the attempt to see into the unconscious by a conscious plan, the dream analysis, his natural history of the couch—is directed downward or backward toward that place where the original in individual consciousness connects to the fundamental or foundational in the rules of the human species (Foucault 1970). That is our urge here: we, too, are seeking foundations; we were just less explicit about our own eschatology. For Freud, the conservative, the urge was to make the therapeutic triumph. We were less explicit about what we are doing here, but the analytical impulse, the conditions of our searching, are much the same. Freud had read Darwin. So have we, and we all see ourselves inside, inside time.

The sciences of human beings exist on the assumption that humankind can be known by finding a foundation in our present. Every study in this collection makes this assumption. And if there is a failing to be found in the "evidence" presented, if below the light of these "facts" there is a shadow, a question of knowing the trivial from the profound, the spurious from the significant, the rigorous from its appearance—then it is the epistemological flaw of that assumption. Unfortunately, in this indeterminacy lie all the issues of representation in the human sciences: the fact of an unrelenting, inescapable relativity, the localization of theory to specific topics, regional concerns, the dilemma of hypothesis autonomy—the problem of model or phenomenon specification. And with those come the dangers of fragmentation.

Now, obviously, a human being is an animal. All human sciences know this. Anthropology, almost from the very beginning, has sought to elucidate the relationship between nature and culture. Sociology, too, has been through the nature-nurture debate. (It was not that long ago that social Darwinism was accepted and taught.) And, as every freshman knows, psychology has long debated the origin of madness, some arguing for somatic disturbances, others for cultural or interactional genesis. But none of this history has put the issues to rest, or changed many of the conditions for theorizing.

So the old paradoxes remain. On the one hand it is argued that human beings are governed by their systems, by their language, their society, their biology. In modern thought it is possible to have access to human "nature" only by means of human words, work, organism. They are the finite limits of our being, if somewhat behind our conscious back. Yet, on the other hand, the thinking that posits these limits is always—so we say— accomplished *within* those conditions. Hence, a constant and threatening ironic danger hangs over all theorizing.[2] Consider, for example, thinking about the workings of the genetic code in humans. Obviously, all thinking that does not claim to be transcendent of the workings of this code is allowed by its workings. The argument must be that humans can no more escape this than they can escape their physical nature. So the old hierarchy still lurks about. In the order of things, physics come first, then physical chemistry, then chemistry, biochemistry, biology—the nesting of Russian dolls in Russian dolls, the emergence of greater systems dependent on more basic systems, each following the rules of the one below. And so, also lurking about is the tendency to automatically attach the human sciences. If one drops any social actor, person with a personality, worker, social sciences from any height he follows the laws of physics right down.

But consider the other side. In Jacob's words: "Of all living organisms, man has the most open and flexible genetic program" (1973, p. 321). Yes, but is it open enough to allow us to produce a biology, a science of gene cloning in which, or from which, there may come a time when DNA allows human reason to produce a science independent enough to badly foul the gene pool in a doomsday holocaust? Jacob says, "With the accumulation of knowledge, man has become the first product of evolution capable of controlling evolution" (1973, p. 322).

The operative word, of course, is *control*. So long as the approach to social use of genetic biology remains piecemeal, so long as the untangling of the program proceeds by "head-down" science, a finding here, a failure there, a success—even a patented success—the human control of biology will be an open question. There remains the possibility of a fundamental error; there is, too, the possible hostile use of biology just as there is with nuclear physics and chemistry. Like many others, Jacob argues that "There is a coherence in the descriptions of science, a unity in its explanations, that reflects an underlying unity in the entities and principles involved" (1973, p. 322). But even if that is true, the use of science lacks such a coherence. The important questions, it would seem, lie not with the limits, the inflexibilities, of our biology or personality or society, but instead, with their flexibilities, with the choices our biology, personalities, societies allow. It is too easy to say that natural selection no longer plays the leading role in transforming man. We do not know what selection looks like when it comes to "natural" science or to the naturalness of science. Perhaps the bomb will in fact make an end or a massive rupture in human DNA. There is that evidence to consider, too. Maybe the center will not hold. All we know is that the bets are down.

That may seem a harsh mention, one notably absent from academic discussions, but it still remains the fact that atomic bombs have already had an effect on the human gene pool. And a very good case can be made that the hydrogen bomb has an effect on the human psyche. There is a sense of being late, getting later. Modeling the relationships among human biology, personality (or mind), and society without accounting for the presence of this factor that in thirty minutes can radically alter the human gene pool, seriously rupture society, and massively "freak out" personalities is the search for wholeness without including all things.[3] Which is, in one sense, the point; in another, not. Even if we ignore the state of the world while studying the state of the world, there is a recursivity to acknowledge here, a certain turning back of things upon themselves that has to be considered. One simply cannot stack sociology or cultural anthropology or psychology or economics on top of the other sciences. The "sciences" of man do not fit in the "system of sciences" that way. Sociology is not about to be the "queen" of sciences. Neither are any of the rest.

So what are we left with? What is there in the issue of evidence? Anything other than something to think about? It is doubtful, but then, that was our point.

Obviously man is, at once, an animal, a personality, a social being. There never is a separation. We all know it. Every theory of human being, of lived experience, of the social, the cultural, the psychological, the biological (the chemical, the physical) which ignores this will be beleaguered by:

Questions of the separation of facts from values.

Questions drawn from epistemology, as there is a dangerous proximity of philosophy to these matters—just as there is a certain anthropological threat to philosophy, and a certain shielding of hard science from epistemological self-criticism in treating of these issues as considerations for a philosophy of science rather than one of knowledge.

Issues concerning the relation of thinking to being (knowledge for what?)

Threats from an awareness of the power of language, its relations to desire, social power, thought, things.

Questions concerning what is absolute, nomothetic; what is not. That is, questions concerning the relations of the school figures to free skating, human nature, human freedom.

Questions of the possibility of a fundamental error in theorizing, in attributing profundity to findings, "evidence," or in the forms of being positive, or in the society itself.

Questions concerning the nature of change. Questions of history.

But then, if there is indication of something beyond in these attempts here to cross the lacuna between mind and body, nature and nurture, the natural and artificial, without slipping into an easy new naturalism, a sociobiology or a biopsychology, it is only a twinkle, something to think about, argue with, evidence for the forebrain. That was our intention. Beginnings, even if they do not look back, have to start someplace, and for some reason.

Notes

1. This statement should not be interpreted as indicating that biological, psychological, and social components need to be included in all or even most studies (in fact, we would argue the contrary position), but rather, that before we are going to understand the phenomenon of human behavior, we need to integrate these various components.

2. Including theorizing with hypothesis, thus all data. Data are not naturally appearing, after all. The facts are not "just there." Even were they, we select them.

3. Perhaps, then, fragmentation is a successful tactic in avoiding the complicity of science (or what? reason?) in these things.

References

American Psychiatric Association. 1980. *Diagnostic and Statistical Manual of Mental Disorders.* 3rd ed. (DSM III). Washington, D.C.: American Psychiatric Association.

Back, Kurt, and Kenneth Gergen. 1966. Cognitive and motivational factors in aging and disengagement. Pp. 289–295 in Ida Simpson and John McKinney (eds.), *Social Aspects of Aging.* Durham, N.C.: Duke University Press.

Booth, Alan and John Edwards. 1977. Crowding and human sexual behavior. *Social Forces* 55(3):791–807.

Brim, Orville. 1968. "Adult Socialization." In John Clausen (ed.), *Socialization and Society.* Boston: Little, Brown.

Brocki, Severine. 1979. Marital status, sex and mental well-being. Unpublished Ph.D. dissertation, Vanderbilt University.

Campbell, Angus, Philip Converse, and Willard Rodgers. 1976. *The Quality of American Life.* New York: Russell Sage Foundation.

Crutchfield, Robert. 1980. An examination of the demographic, mental health, and social role correlates of adult drug use. Unpublished Ph.D. dissertation, Vanderbilt University.

Cumming, Elaine, and William Henry. 1961. *Growing Old: The Process of Disengagement.* New York: Basic Books.

Dean, Lois R. 1962. The human life cycle. Pp. 286–292 in *International Encyclopedia of the Social Sciences.* New York: Macmillan and the Free Press.

———. 1963. *Childhood and Society.* New York: Norton.

Erickson, Eric. 1963. *Childhood and Society.* New York: Norton.

———. 1968. *Identity: Youth and Crisis.* New York: Norton.

Finch, Caleb, and Leonard Heyflick. 1973. *Handbook of the Biology of Aging.* New York: Van Nostrand.

Foucault, Michel. 1970. *The Order of Things.* New York: Pantheon.

Foner, Anne, and David Kertzer. 1978. Transition over the life course: Lessons from age-set societies. *American Journal of Sociology* 83 (March):1081–1104.

Gaitz, Charles, and Judith Scott. 1972. Age and the measurement of mental health. *Journal of Health and Social Behavior* 13(March):55–67.

Gordon, Chad. 1972. Role and value development across the life-cycle. Pp. 65–106 in J.A. Jackson (ed.), *Role.* Cambridge: Bainbridge University Press.

Helson, H. 1964. *Adaptation-Level Theory: An Experimental and Systematic Approach to Behavior.* New York: Harper.

Jacob, Francois. 1973. *The Logic of Life.* New York: Pantheon.

Jarvis, George, Robert Ferrence, F. Gordon Johnson, and Paul Whitehead.

1976. Sexual age patterns of self-injury. *Journal of Health and Social Behavior* 17(June):146–155.

Jung, Carl. 1963. *Memories, Dreams, Reflections.* New York: Pantheon.

———. 1964. *Man and His Symbols.* Garden City, N.J.: Doubleday.

Kuhlen, Raymond G. 1959. Aging and life adjustment. Pp. 852–897 in J. Birren (ed.), *Handbook of Aging and the Individual.* Chicago: University Press.

Levinson, Daniel, Charlotte Darrow, Edward Klien, Maria Levinson, and Braxton McKee. 1978. *The Seasons of a Man's Life.* New York: Ballantine.

Lidz, Theodore. 1968. *The Person: His Development through the Life Cycle.* New York: Basic Books.

Lowenthal, Marjorie Fiske, Majdu Thurnber, and David Chiriboga. 1975. *Four Seasons of Life.* San Francisco: Jossey-Bass.

McAucliffe, William. 1975. Beyond secondary deviance: Negative labelling and its effects on the heroin addict. Pp. 205–242 in Walter Gove (ed.), *The Labelling of Deviance: Evaluating a Perspective.* New York: Sage/Halstead.

Mellinger, G., M. Balter, D. Manheim, I. Cisin, and H. Parry. 1978. Psychic distress, life crises and the use of psychotherapeutic medications: National household survey data. *Archives of General Psychiatry* 35 September):1045–1052.

Neugarten, Bernice. 1968. The awareness of middle age. Pp. 93–98 in Bernice Neugarten (ed.) *Middle Age and Aging: A Reader in Social Psychology.* Chicago: University of Chicago Press.

Pressey, Sidney, and Raymond Kuhlen. 1957. *Psychological Development through the Life Span.* New York: Harper and Brothers.

Robbins, Lee. 1975. Alcoholism and labelling theory. Pp. 21–34 in Walter Gove (ed.), *The Labelling of Deviance: Evaluating a Perspective.* New York: Sage/Halsted.

Rowe, Alan, and Charles Tittle. 1977. Life cycle changes and criminal propensity. *The Sociological Quarterly* 18(Spring):223–236.

Sears, Robert, and S. Shirley Feldman (eds.). 1964. *The Seven Ages of Man.* Los Altos, Calif.: William Kaufmann.

Spirluso, Waneen Wyrick. 1980. Physical fitness, aging, and psychomotor speed: A review. *Journal of Gerontology* 35(November):850–865.

Uniform Crime Reports. 1971. *Crime in the United States.* Washington, D.C.: U.S. Government Printing Office.

Index

About the Contributors

M.S. Buchsbaum, M.D., is chief of the Section on Clinical Psychophysiology, Biological Psychiatry Branch, National Institutes of Mental Health.

W.E. Bunney, Jr., M.D., is head of the Biological Psychiatry Branch, National Institutes of Mental Health.

Linnda Caporael, Ph.D., is assistant professor of psychology, Department of Psychology, Renselaer Polytechnic Institute. She received the Ph.D. in psychology from the University of California at Santa Barbara.

L.L. Cavalli-Sforza was a microbial geneticist in the first decade of research in that field and later a student of human population genetics with a special interest in human variation and evolution, including cultural change. He is professor of genetics at Stanford Medical Center, Stanford, California.

Juan B. Cortés, Ph.D., S.J., is a professor of psychology at Georgetown University. He received the Ph.D. in clinical psychology from Harvard University. He is the author of numerous articles on psychology and of three books.

G.C. Davis, M.D., is associate of psychiatry care, Western Reserve School of Medicine, and director of Psychiatry Research Program, Cleveland Veterans Administration Hospital. He received the M.D. from Duke University.

C. Dawn Delozier, M.S., is assistant in medical genetics at the Institut de Génétique Médicale, Hôspital Cantonel, Geneve, Switzerland.

Eric Engle, M.D., is director of the Institut de Génétique Médicale, Hôspital Cantonel, Geneve, Switzerland.

Marcus Feldman, Ph.D., is professor and associate chairman in the Department of Biological Sciences, Stanford University. His research interests include population genetics of insects, marine organisms, and man, with a special focus on evolutionary theory and interpretation of genetic variability.

Kenneth K. Kidd received the Ph. D. in genetics from the University of Wisconsin, Madison. He currently is associate professor of human genetics and psychiatry at Yale University School of Medicine. His research has included studies of human evolution and is now focused on the genetics of several complex human disorders, particularly behavioral disorders.

Gerald L. Klerman, M.D., is director of research, Stanley Cobb Psychiatric Research Laboratories, Massachusetts General Hospital, and professor of psychiatry, Harvard University. He is the former administrator of the Alcohol, Drug Abuse and Mental Health Administration.

Jacquelynne Eccles Parsons received the Ph.D. in developmental psychology from the University of California at Los Angeles in 1971. She is currently associate professor of psychology at the University of Michigan, Ann Arbor. She is editor of *The Psychosociology of Sex Differences and Sex Roles* (1980) and is coauthor of *Women and Sex Roles* (1978).

Gail Schechter worked for a number of years in the Intramural Research Program of the National Institute of Mental Health, Bethesda, Maryland, doing biobehavioral studies of sensory processing in psychiatric and normal populations. Subsequently, she assisted Dr. Gerald Klerman in the Alcohol, Drug Abuse and Mental Health Administration with special projects in the field of mental health, especially the interactions between pharmacological and psychological treatments of mental disorders. Currently she is a doctoral student in psychology at the University of California, San Francisco.

Paul Taubman is professor of economics and chairman of the Economics Department at the University of Pennsylvania. He is a Fellow of the Econometric Society and the International Society of Twin Studies. He has published over fifty books and articles.

Ingrid Waldron, Ph.D., is associate professor of biology at the University of Pennsylvania. Her research has analyzed causes of sex differences in mortality and morbidity, the relationships between employment and women's health, and causes of cross-cultural variation in blood pressure and serum cholesterol.

Herbert Weiner, M.D., is professor of psychiatry and neuroscience, Albert Einstein College of Medicine; chairman of the Department of Psychiatry, Montefiore Hospital and Medical Center, New York; and editor-in-chief, *Psychosomatic Medicine.* He has been president of the American Psychosomatic Society and the Association for Research in Nervous and Mental Disease.

About the Editors

Walter R. Gove received the B.S. from New York University College of Forestry and the M.A. and Ph.D. in sociology from the University of Washington, Seattle. He is currently a professor of sociology at Vanderbilt University. He is the author of numerous articles on mental illness, crime, drug use, sex roles, the family, and the effects of living arrangements. He is the editor of *Labelling Deviant Behavior* and coauthor of *Household Overcrowding* and *Psychological Consequences of Living Alone*.

G. Russell Carpenter received the Ph.D. in sociology from Indiana University. He has taught at Indiana and Vanderbilt Universities. He is currently writing articles on popular subjects in sociology.